Lecture Notes in Economics
and Mathematical Systems

504

Springer
Berlin
Heidelberg
New York
Barcelona
Hong Kong
London
Milan
Paris
Singapore
Tokyo

Pierre-Yves Moix

The Measurement of Market Risk

Modelling of Risk Factors, Asset Pricing,
and Approximation of Portfolio Distributions

 Springer

332.601
M71m

Author

Dr. Pierre-Yves Moix, CFA
Untere Torfeldstrasse 52
5033 Buchs/AG
Switzerland

Cataloging-in-Publication data applied for

Die Deutsche Bibliothek - CIP-Einheitsaufnahme

Moix, Pierre-Yves:
The measurement of market risk : modelling of risk factors, asset
pricing, and approximation of portfolio distributions / Pierre-Yves
Moix. - Berlin ; Heidelberg ; New York ; Barcelona ; Hong Kong ;
London ; Milan ; Paris ; Singapore ; Tokyo : Springer, 2001
  (Lecture notes in economics and mathematical systems ; 504)
  ISBN 3-540-42143-2

ISSN 0075-8450
ISBN 3-540-42143-2 Springer-Verlag Berlin Heidelberg New York

Springer-Verlag Berlin Heidelberg New York
a member of BertelsmannSpringer Science+Business Media GmbH

http://www.springer.de

© Springer-Verlag Berlin Heidelberg 2001
Printed in Germany

Typesetting: Camera ready by author/editors
Cover design: *design & production,* Heidelberg

Printed on acid-free paper   SPIN: 10840038   55/3142/du   5 4 3 2 1 0

To

Sophie Rose, Susanne,
Marie-Rose and Gérard

To the memory of my oncle
Ernest Moix
(1927-1999)

"L'absence est à l'amour ce qu'est au feu le vent;
il éteint le petit et allume le grand"

Roger de Bussy-Rabuttin
"Histoire amoureuse de Gaules"

Preface

This book is a revised version of my doctoral dissertation submitted to the University of St. Gallen in October 1999. I would like to thank Dr.oec. Marc Wildi whose careful reading of much of the text led to many improvements. All errors remain mine.

Pfäffikon SZ, Switzerland, March 2001   Pierre-Yves Moix

Preface to the dissertation

> *"Education is man's going forward from cocksure ignorance to thoughtful uncertainty"*

Don Clark's Scrapbook quoted in Wonnacott and Wonnacott (1990).

After several years of banking practice, I decided to give up some of my certitudes and considered this thesis project a good opportunity to study some of the quantitative tools necessary for the modelling of uncertainty. I owe very much to Prof. Dr. Karl Frauendorfer, the referee of my thesis, for the time he took to read the manuscript and for the numerous valuable suggestions he made. I am also very grateful to Prof. Dr. Klaus Spremann who kindly accepted to co-refer my thesis and who strengthened my interest in finance during my study period. During my time at the Institute for Operations Research of the University of St. Gallen (IfU-HSG) I had the opportunity to participate in the project "RiskLab" which provides a very profitable link between finance practice and academics. I would especially like to thank Dr. Christophe Rouvinez from Credit Suisse for his comments and all the data he provided so generously. Likewise, I am very grateful to my colleagues at the institute who have been a valuable source of new ideas. I especially profited from Dr.oec. Christina Mahron's programming skills, from the mathematical expertise of Dipl.-Math. Veronika Halder, and from the numerous discussions with Dr.oec. Ralf Gaese, Dipl.-Math.oec. Guido Haarbrücker, and Dipl.-Wirtsch.Ing. Stefan Scholz. Many thanks to Claudia Rossi-Mayer, CELTA, who spent many days at correcting the manuscript and improving my modest knowledge of English. Last but not least, my warmest thanks go to my entire family for all the love and support they have given me, and above all for their patience.

Contents

1. Introduction

1.1 The Need for Risk Measurement

Financial risk, the financial exposure to uncertainty, has become a crucial issue for both financial and non-financial corporations. Recent scandals (Barings, Orange County, Metallgesellschaft, LTCM) show the need for powerful risk management tools. In addition to the recent debacles, several factors have contributed to the development of a consistent and integrated treatment of financial risk:

- the increasing volatility of the markets
- the development of information technology
- the deregulation of financial markets
- the processes of securisation and disintermediation
- the shift in the legal and regulatory environment.

It is generally accepted that the financial world has become riskier over the past decades. While the volatility of the equity markets has been observed at least since the first crash of 1929, other markets have experienced an increase of uncertainty more recently. The breakdown of the Bretton Woods agreement has led to an evident increase in exchange rate volatility. As the central banks began to target money supply growth to restrain inflation, interest rates in turn have become more volatile. Finally, the price of several basic commodities, especially petroleum, began to fluctuate heavily in the 70's.

The general increase in volatility of the financial markets can be partly explained by the growing uncertainty of the economic environment. It is, on the other hand, a reflection of the growing efficiency and integration of the financial markets. The development of information technology and the integration of financial markets allow the almost instantaneous move of capital from one market place to another and therefore the rapid assimilation of any new, available information. In efficient markets, all current information, which is supposed to arrive at random, is fully reflected in the prices. As a result, future prices are hard to predict and uncertainty increases with the time horizon. Hence forecasting cannot achieve any degree of accuracy and consequently cannot eliminate price risk. However, the probabilistic modelling of the price dynamics allows to measure and therefore to manage financial risk.

Further, and as a consequence of the growing volatility, the financial markets have been deregulated relatively recently. Protective barriers have been removed and free circulation of capital has been facilitated. Most financial institutions have lost their secure earnings and have increasingly been exposed to competition. Simultaneously, new profit opportunities have appeared for market participants with sufficient analytical skills for the development of sophisticated products. Particularly the derivatives market has experienced a rapid growth reflecting the advances of financial theory after the pioneering work of Black and Scholes. The widespread use of derivatives, particularly for speculative aims, has given rise to an intense debate and to increasing concerns. What is often overlooked in this context is that derivatives are in fact equivalent to a combination of existing financial instruments and thus do not create new risks. Moreover, they facilitate the transfer and the synthesisation of risks and are therefore ideal risk management tools. Nevertheless, the almost unlimited potential of leverage and the non-linear exposure to movements in the underlying markets make the use of derivatives challenging. Derivatives have further broken down the boundaries between the various asset classes. Equity-Index-Linked Swaps for example convert the interest rate exposure to cash flow tied to the return to an equity index.

In addition to the development of the derivatives market the general trend towards securisation and disintermediation has also contributed to the development of risk management. Securisation, the conversion of traditional bank financing into tradeable obligations, requires the periodic evaluation of the risks associated with these obligations. As a consequence, marking to market practices have gradually replaced traditional accrual accounting. The process of disintermediation leads top rated non-financial institutions to turn more and more directly to capital markets. Thus, these corporates have to gain the necessary expertise in risk management. A side-effect of disintermediation is that banks suffer from a reduction in tier-1 loan customers.

The regulatory pressure has also played an important role in the development of new risk management methodologies. In 1995, for instance, the Basle Committee on Banking Supervision issued a package of supervisory proposals for applying capital charges to the market risks incurred by banks (see BIS (1995c,b,a)). At first the committee developed a standardised measurement framework and then it proposed to allow financial institutions to use their in-house models as an alternative. The Basle rules have applied since the beginning of 1998 to banks in the G10 countries for the positions in trading books and all foreign exchange exposures. The firms that wish to use proprietary models have to meet a number of quantitative and qualitative criteria. The quantitative criteria suggest the use of Value-at-Risk for determining capital charges.

1.2 The Nature of Financial Risk

Financial risk is generally divided into the following categories:

- *Market risk*: the risk resulting from changes in financial asset prices. Market risk can be further disaggregated by the exposures to the *risk factors*, for instance equity and commodity prices, as well as interest rates and foreign exchange rates.
- *Credit risk*: the risk arising from credit events like default, upgrade or downgrade of a counterparty.
- *Liquidity risk*: the risk arising from the impossibility to buy or sell assets quickly or at the prevailing market rates because of the lack of sufficient market activity.

The exposure to financial risk is obvious and a byproduct of the core business of financial institutions. Most of their assets and liabilities are of a financial nature and it is the primary task of risk management to evaluate and control the residual net risk remaining after the matching of all positions. Bank's assets and liabilities are often split into the trading and the banking book. The trading book contains all positions held for the purpose of short term dealing. Most positions are traded on liquid markets. The trading book is exposed to all forms of market risks. The banking book includes deposits, loans, and the investment portfolio of the bank. The banking book is especially sensitive to changes in the level of interest rates and more exposed to liquidity risk than the trading book. For global players risk taking has become very decentralised since trading units are localised all around the world. Moreover, traders often have an incentive to take maximum risk because of the asymmetric profile of their compensation schemes.

By contrast non-financial institutions have substantial positions such as plant, equipment and real estate which are not of a financial nature and difficult to value. Moreover, a large part of their financial risks arises from the non-financial core business. For example corporates involved in international trade are faced with exchange rate risks. As a result, financial risk is perpetual and difficult to hedge with financial products of finite maturity. Financial risk is, however, not only a side-effect of business activity but may also have real effects on the competitive exposure of non-financial institutions. Changes in exchange rates can become an advantage for foreign competitors and fluctuations in commodity prices can increase the attractiveness of substitute products. Moreover, exchange and interest rates as well as commodity prices are key factors affecting the value of investment opportunities. The primary task of risk management for non-financial institutions consists in supporting investment decisions through management of the cash flows, i.e. through generation of the cash necessary to fund investments (Froot et al. (1994) is an excellent article on the subject).

Note that there is a large debate about the value of financial risk management for publicly held companies. Modigliani and Miller (1958) have shown that in a world with perfect information and without taxes and transaction costs, the debt-equity structure of a corporation does not matter. Moreover, portfolio theory suggests that a firm has no incentive to hedge away any kind of financial risk since investors can eliminate the firm-specific risk at a much lower cost through diversification of their portfolio holdings. There is therefore no reason to reduce risk on behalf of the investors. However, in the real world risk management can increase the value of a company by reducing taxes and/or avoiding transaction costs that go together with a lowering of the credit quality or a financial distress. More importantly, as discussed above, risk management can adapt the financial exposure of a company to its needs for value-enhancing investments. The challenge of risk management is to develop adequate approaches to and strategies for the specific needs of companies, i.e. to be consistent with the companies' business strategies.

Investors, especially institutional investors, such as pension and mutual funds or insurances, also have to monitor and control the financial risk they are exposed to. The need for risk management tools is increasingly important as institutional investors turn to international markets and invest in complex instruments in order to face the pressure for returns from the shareholders, beneficiaries, or policyholders. The liabilities of institutional investors are often linked to actuarial values, such as retirement statistics, mortality rates, or the occurrence of casualties like fire or earthquakes. What is of main interest for insurance companies and pension funds is the management of the surplus, i.e. the development of a strategy enhancing the probability that the liabilities can be paid from the assets. Mutual fund risk is generally relative to some benchmark which represents a feasible investment alternative. The skill of the fund managers is measured by comparing the risk-adjusted returns of the fund with those of its benchmark.

The prerequisite for successful risk management is careful risk measurement. The issue of measurement of market risk is the central theme of this book. The quantitative methods discussed herein are, however, general enough to incorporate the modelling of credit and liquidity risks. On the other hand, legal and operational risks, which also play a central role in the practice of risk management, are much more difficult to quantify and are generally better managed with qualitative tools. It is worth noting that a successful risk management requires the creation of an appropriate corporate risk culture and risk education.

Quantitative risk measurement encompasses several dimensions which have been described by Lo (1999) as the three P's: preferences, probabilities and pricing. The preferences reflect the behaviour of economic agents when faced with uncertainty about the future. Probabilities represent a formal description of uncertainty by assessing the likelihood of future events. The risk factors are the source of uncertainty in the financial market and their dynamics are modelled in probabilistic terms (the *distribution information*). Pricing relies on preferences and probabilities to determine the exact value of financial assets as a function of the underlying risk factors.

The methodology discussed in this book supports an integrated view of risk measurement. It is suited to all risk factors and asset classes and takes into account the three P's. Moreover, it makes allowance for the interactions between the risk factors and therefore includes the portfolio (diversification) effects. The use of an integrated approach has the further advantage of allowing the development of unified rules and the promotion of firm-wide risk management and culture. Traditional risk measurement, on the other hand, is generally incomplete, neglecting either the distribution or the value information, and is specific to a particular asset class. As an example the duration/convexity analysis has been developed exclusively for interest-rate sensitive securities and assesses the risk arising from maturity mismatches. Duration/convexity analysis gives an estimate of the sensitivity of financial instruments to an arbitrary change in the interest rates. However, no attempt is made in this context to consider the distribution information, i.e. to model the dynamics of the interest rates.

1.3 Formal Framework

It is the objective of this book to set up a quantitative economic model for the assessment of market risks. Any model involves abstraction, i.e. the simplification of the real world in order to construct a mathematically tractable model.

Models have proven to be very successful in the field of engineering since they can adequately capture the relationships in the physical world. Economic models on the other hand involve a formalisation of human (economic) behaviour, which is definitely, and fortunately, too rich to be fully described in quantitative terms. It is generally assumed that the agents of an economy can be studied in isolation and that their primary motive is the maximisation of their monetary wealth. Furthermore, the individuals are supposed to be non-satiated, i.e. greedy. Moreover, they behave rationally: faced with uncertainty they do not behave in an arbitrary fashion but pursue a clear objective, which is the maximisation of their *utility*. Finally, they are able to process all available information.

Clearly, this description of human behavior may appear very simplistic. In particular the hypothesis of *rationality*, whose main advantage lies in its parsimonious representation of human behaviour, is subject to growing criticism. *Behavioural finance*, which will be briefly discussed in Sect. 2.1.1, proposes alternative models. However, as stated above, no theory will be able to fully codify human behaviour. It is therefore important to keep the limitations of any economic model in mind and to be aware of the *model risk*, the risk that simplifying assumptions produce inaccurate results. A model can not replace the expertise and judgment of professionals. Modelling, however, remains an efficient way of analysing systematically the important interrelationships involved in a problem.

This section provides the formal framework for the rest of the work. It will be assumed that a *financial decision*, for example the building of a portfolio of financial assets, is made at some initial date $t = 0$. The (monetary) *consequence* of the decision, e.g. the value of the portfolio, will be measured for some terminal date T with $T > 0$. The consequence is a function of *decision variables*, e.g. the weights of the different assets in the portfolio, and of *empirical variables*, e.g. *risk factors* which determine the value of the assets at time T. The decision-maker exercises direct control over the decision variables but has virtually no influence on the outcome of the empirical variables, whose values arise from decisions made by many individual agents. As a result, the consequence is not known for certain at time $t = 0$ and the decision has to be made under uncertainty. *Decision-making under uncertainty* involves the assessment of the probable consequences, which requires a quantitative modelling of the uncertainty represented by risk factors.

1.3.1 Modelling the Uncertainty

Probability is the most widely used formalism to describe uncertainty (see appendix A for a more detailed discussion). The probabilistic representation of uncertainty is given by a set of possible *states of the world* to be realised at time T and whose outcomes are denoted by ω. Only the states of the world which can be distinguished by their impact on the risk factors are of interest. The outcomes ω are mutually exclusive and collectively exhaustive. Thus, two states can not occur simultaneously and the sum of the probabilities of the individual states equals one.

The *sample space* or *risk factor space* Ω represents the set of all ω. At time $t = 0$ individuals know that the true state of the world is an element of Ω but do not know which state will occur at time T. Note that Ω may have uncountably infinite elements, therefore a probability is generally not defined on Ω but rather on a collection $\mathcal{F}$ of subsets of Ω that satisfies a certain structure. A subset A of Ω is generally called an *event*.

Formally, a collection $\mathcal{F}$ of events which satisfies (see Brémaud (1994), p.4):

1. $\Omega \in \mathcal{F}$,
2. if $A \in \mathcal{F}$ then its complementary $A^c \in \mathcal{F}$,
3. if $A_i \in \mathcal{F}$ is a countable sequence of sets then $\cup_i A_i \in \mathcal{F}$,

is referred to as a *σ-field* or *σ-algebra* on Ω.

Given a family $\mathcal{A}$ of sets in Ω, the smallest σ-algebra containing all sets in $\mathcal{A}$ is called the σ-algebra generated by $\mathcal{A}$. In particular, the σ-algebra generated by the open subsets of $\mathbb{R}^M$ is called the Borel σ-algebra $\mathcal{B}$ and its members are referred to as Borel sets.

$(\Omega, \mathcal{F})$ is called a *measurable space*, i.e. it is a space on which one may put a measure. A measure $z : \mathcal{F} \to \mathbb{R}$ is a non-negative countably additive set function (see Durrett (1996), p.2):

1. $z(A) \geq z(\emptyset) = 0$ for all $A \in \mathcal{F}$ and
2. $z(\cup_i A_i) = \sum_i z(A_i)$ if $A_i \in \mathcal{F}$ is a countable sequence of disjoint sets.

If additionally $z(\Omega) = 1$, z is termed a *probability measure* and will be denoted by P. $P: \mathcal{F} \to [0,1]$ is therefore a function that assigns probabilities to events. The triple $(\Omega, \mathcal{F}, P)$ is referred to as a *probability space*.

This formal description of the probabilistic model allows for modelling the values of the risk factors at some time $t > 0$ in terms of random variables. Formally, a real valued function X defined on Ω is said to be a *random variable* if for every Borel set $B \in \mathcal{B}$:

$$X^{-1}(B) = \{\omega : X(\omega) \in B\} \in \mathcal{F} \tag{1.1}$$

X is said to be $\mathcal{F}$-measurable and induces a probability measure on $\mathbb{R}$ called its distribution.

The distribution of X is usually represented by its *(cumulative) distribution function* $F_X(x) = P(X \leq x)$ with following properties (see Durrett (1996), p.4):

1. F is non-decreasing,
2. $\lim_{x \to \infty} F_X(x) = 1$, $\lim_{x \to -\infty} F_X(x) = 0$,
3. F is right continuous, i.e. $\lim_{y \downarrow x} F_X(y) = F_X(x)$,
4. If $F_X(x_-) = \lim_{y \uparrow x} F_X(y)$ then $F_X(x_-) = P(X < x)$,
5. $P(X = x) = F_X(x) - F_X(x_-)$.

When the distribution function $F_X(x) = P(X \leq x)$ has following representation:

$$F_X(x) = \int_{-\infty}^{x} f_X(y)dy \qquad (1.2)$$

f_X is called the *density function* of X.

The *moments* of X represent a widely used partial representation of the distribution F_X. The moment of order n of X, $E[X^n]$, if it exists, is given with (see for instance Roger (1991), p.76) :

$$E[X^n] = \int_{\Omega} X^n dP = \int_{-\infty}^{+\infty} x^n dF_X(x). \qquad (1.3)$$

The first moment is called the mean or expected value is usually denoted by μ. The *central moment* μ_n of order n, provided it exists, is defined as:

$$\mu_n = E\left[(x - \mu)^n\right] = \int_{\Omega} (X - \mu)^n dP. \qquad (1.4)$$

$\mu_2 = \text{var}[X]$ is called the *variance* of X and gives information about the dispersion of the distribution around μ. Thus, the variance represents a *scale parameter* of the distribution. The square root of the variance is called the *standard deviation* of the distribution. μ_3 gives insight into the asymmetry of the distribution about its mean, which is also called *skewness*. The fourth central moment provides information about the relative weight of the centre of the distribution, which is referred to as *kurtosis*.

A random variable defined on the probability space $(\Omega, \mathcal{F}, P)$ and assuming values in $(\mathbb{R}^M, \mathcal{B})$, $M \in \mathbb{N}$ and $M > 1$, is also called a *random vector*. M represents the dimension of the vector. Note that in our model the consequences are measured for the single time $t = T$ and therefore only the distribution of the risk factors at time T is of interest. It will, however, be shown in Chap. 3 that the adequate modelling of the distribution of the risk factors at time T requires the analysis of their evolution through time. Formally, a collection of random variables $\{X(t)\}$, where $t \in \mathcal{T}$ represents the time, defined on the probability space $(\Omega, \mathcal{F}, P)$ and assuming values in $(\mathbb{R}^M, \mathcal{B})$, $M \in \mathbb{N}$, is called a *stochastic process*. If $M > 1$ $\{X(t)\}$ is also called a vector process.

1.3.2 The Information Structure

Let's consider again the probability space $(\Omega, \mathcal{F}, P)$. $\mathcal{F}$ represents the set of events A for which a probability $P(A)$ can be associated. The assignment of probability to events is done on the basis of the information available about the states of the world at a given point in time t. The σ-algebra on Ω is therefore a function of time and is more precisely denoted by $\mathcal{F}(t)$. With the passage of time new information is being gathered and it is assumed that past information is not forgotten. As a result, the knowledge about the states of the world increases as time progresses and $\mathcal{F}(t) \subseteq \mathcal{F}(s)$ for $s > t$. The non-decreasing sequence of σ-algebras $\mathbb{F} = \{\mathcal{F}(t), t \in \mathcal{T}\}$ is called a *filtration*. A stochastic process $\{X(t)\}$ is said to be *adapted* to the filtration (see Roger (1991), p.172) if for any t, $X(t)$ is $\mathcal{F}(t)$-measurable, i.e. for any time $t \in \mathcal{T}$ $X(t)$ is a random variable on $(\Omega, \mathcal{F}(t))$. In other words, if $\{X(t)\}$ is $\mathbb{F}$-adapted and an agent's information filtration is $\mathbb{F}$, then the outcome of $X(t)$ is known (to the agent) at time t. Note that if Ω is a finite set of states, each σ-algebra $\mathcal{F}(t)$ is generated by a family $A'_i, (i = 1, \cdots, N)$ of subsets of Ω called a *partition* $\mathcal{P}(t)$ of Ω which exhibits following properties (see Rebonato (1998), p.453):

1. the intersection of any two elements A_i and A_j of $\mathcal{P}(t)$ is empty:
   $A'_i \cap A'_j = \emptyset \qquad i, j = 1, \cdots, N, \qquad i \neq j,$
2. the union of all elements of $\mathcal{P}(t)$ gives the sample space:
   $A'_1 \cup A'_2 \cup \cdots \cup A_N = \Omega.$

A partition is said to be finer to another if any event in the latter partition represents the union of some events in the former. Clearly, with the assumption of non-decreasing σ-algebras the partition of Ω will be finer with the passage of time.

1.4 Problem Statement

Consider an economy where agents, individuals or firms, may exchange in a market I financial instruments which are traded in home currency as numéraire. The financial instruments are exposed to the changes in value of a set of M risk factors also termed market rates or markets prices. A typical set of risk factors may include the prices of stocks, commodities, foreign currencies and the level of interest rates for the currencies of interest. The risk factors are represented by a stochastic process $w_i(t), i = 1, 2, \cdots, M$ which is defined on the probability space $(\Omega, \mathcal{F}, P)$. At time $t = 0$, $w(T) = (w_1(T), \cdots, w_M(T))$ is a M-dimensional random variable which represents the possible outcomes of the risk factors at time T. Clearly, the current values of the risk factors, the origin, are known and denoted by $w(0) = (w_1(0), w_2(0), \cdots, w_M(0))$. A particular realisation of $w(T)$ is referred to as a *(market) scenario*. The *nominal scenario* w^0 represents a specific value for the value of the risk factors at time T. In the following the nominal scenario will be defined as the expected value of the risk factor value at time T:

$$w^0 = E[w(T)] = \int_\Omega w(T) dP. \qquad (1.5)$$

In principle the value of each financial instrument could be modelled directly as a risk factor. However, this approach would increase the dimensionality of the problem considerably since $I \gg M$. Moreover, the distributional information of new, respectively illiquid, instruments may be not available, respectively inaccurate. A better way of proceeding consists therefore in modelling the price of the financial assets as a function $g()$ of one or several risk factors. Formally, $g_i(w) : \mathbb{R}^M \rightarrow \mathbb{R}$ denotes the real valued *price function (pay-off profile, value function, risk profile)* of a financial instrument $i = 1, 2, \cdots, I$. It will be assumed that g_i is known for every instrument i and gives a value for every outcome of $w(T)$.

At time $t = 0$ the agents can buy and sell financial instruments. The *portfolio position* of an individual agent is given in terms of the vector $\theta = \theta_1, \cdots, \theta_I$, where θ_i denotes the quantity of the ith instrument held at time 0. θ may have positive as well as negative entries, which are called *long positions* or *long holdings*, respectively *short positions* or *short holdings*. During the time interval $]0, T]$, the *holding period* or the *horizon*, no investment can be made so that the positions of the portfolio remain fixed. $]0, T]$ can be a fixed period imposed by a regulator, the period required to liquidate a position or the interval of time between asset allocation or hedging decisions.

The *value function* of the portfolio at time T is denoted by $g[w(T)]$ and is given with:

$$g[w(T)] = \sum_{i=1}^{I} \theta_i g_i[w(T)]. \tag{1.6}$$

Clearly, the value of the portfolio is stochastic. The portfolio distribution is given by the induced probability measure P_g of the transformed random variable $g(w)$ on $(\mathbb{R}, \mathcal{B})$. The associated distribution function F_g is given through:

$$F_g(v) = P\left(g[w(T)] \leq v\right). \tag{1.7}$$

The change in value of the portfolio at time T, also called the *profit and loss (P&L)* function, is denoted by $\bar{g}$ and defined as:

$$\bar{g}[w(T)] = \sum_{i=1}^{I} \theta_i \left[g_i[w(T)] - g_i[w(0)]\right] \tag{1.8}$$

with associated probability measure $P_{\bar{g}}$ and distribution function $F_{\bar{g}}$. Obviously, a negative value of $\bar{g}[w(T)]$ represents a loss and a positive value a profit.

The (simple rate of) *return* of the portfolio over the horizon, $R_g[w(T)]$ is given through:

$$R_g[w(T)] = \frac{g[w(T)] - g[w(0)]}{g[w(0)]}. \tag{1.9}$$

It associated probability measure, respectively distribution function, is denoted by P_{Rg}, respectively by F_{Rg}.

The notions of portfolio value, profit and loss and return are obviously directly related. For a given portfolio value $g[w(0)]$, the profit and loss and return distributions can be directly derived from the portfolio distribution.

1.5 Structure of the Book

Chapter 2 reviews the approaches to decision-making under uncertainty where uncertainty is summarised by the portfolio distribution. The individuals face two types of decisions. The first decision, the *portfolio decision*, is the selection of the portfolio at time $t = 0$, i.e. the allocation of a given amount of money among the available investment possibilities. The second decision, the *capital requirement decision*, is the assessment of the amount of economic capital needed to cover the potential losses at time T.

The portfolio decision involves the choice of a portfolio among infinite combinations of financial instruments. This requires rules allowing the ordering of a large number of risky alternatives. Each of them can be characterised by a probability distribution over the possible outcomes. The agents are supposed to be *non-satiated*, i.e. prefer more than less, and obey the axiomatic behaviour defined by von Neumann and Morgenstern. As a result, it is possible to construct a *utility function* u that assigns a value over the possible outcomes of the portfolio distribution. Moreover, the decision criterion of the agent is the maximisation of expected utility. Note that complete information about individual utility functions is seldom available and decision rules for the maximisation of expected utility are based on limited information about u. The most general operationalisation of the expected utility maximisation is the *stochastic dominance* approach which is adequate for entire classes of utility functions. However, the stochastic dominance criterion involves the pairwise comparison of alternative portfolios and is therefore of little use when there is a large number of alternatives. A second approach, the *mean-risk analysis*, works with the separate dimensions of expected value and risk. The approach assumes that a larger expected value is preferred to a smaller one and a smaller risk is preferred to a larger one. Within this framework of the mean-risk analysis several alternative measures $r()$ of risk have been proposed. All measures discussed herein are special cases of the Benell Stone's *generalised risk measure* (see Fishburn (1977)):

$$r(F_{Rg}) = \int_a^{\gamma(F_{Rg})} | \, \eta(F_{Rg}) - x \, |^\alpha \, dF_{Rg}(x) \qquad \alpha \geq 0 \qquad (1.10)$$

where $\eta(F_{Rg})$ is a function of F_{Rg} which determines the reference level from which deviates are measured, $\gamma(F_{Rg})$ determines the range parameter, specifying what deviations are to be included in the risk measure, and α is a measure of the relative impact of large and small deviations.

The most widely used special cases of Stone's risk measure are the *variance* :

$$\text{var} \; = \int_a^b (x - \mu)^2 dF_{Rg}(x), \qquad (1.11)$$

respectively the *lower partial moments* of order n:

$$\text{LPM}_{n,l} = \int_a^l (l - x)^n dF_{Rg}(x), \qquad n \in \mathbb{N}_0, \qquad (1.12)$$

where l is an arbitrary reference level. Unlike standard deviation, lower partial moments only include the deviations below the reference level l. Both the mean-variance and mean-LPM rules are consistent with the expected utility maximisation for restricted classes of utility functions.

The second part of chapter 2 is devoted to the capital requirement decision, which represents the assessment of the worst case outcome of the portfolio, i.e. the potential *maximum loss* occurring at time T. The potential maximum loss is, however, not a very informative risk measure. It is actually easy to construct with short positions a portfolio which is open to unlimited loss. As a result, alternative risk measures capital requirement decision are proposed in the literature. The most well-known, referred to as *Value-at-Risk* (VaR) , is defined as:

$$VaR_\alpha = v_\alpha = - \inf \{x \mid P\left(\bar{g}[w(T)] \leq x\right) \geq \alpha\} \qquad (1.13)$$

Hence, Value-at-Risk (VaR) represents a loss limit which is exceeded with respect to a (small) given frequency α. VaR has become the risk management standard since it has been proposed by the Basle Committee on Banking Supervision for regulatory capital requirement. However, VaR is the issue of a large debate since it fails to satisfy the *subadditivity* property. Section 2.2.2 discusses the class of *coherent risk measures*: which were developed to overcome the conceptual weakness of Value-at-Risk.

Chapters 3 to 5 deal with the modelling of uncertainty, i.e. the approximation of the portfolio distribution. Chapter 3 discusses the *distribution information*, that is the modelling of the risk factor distribution, which represents the source of uncertainty. Chapter 4 is devoted to the *price information*, i.e. the pricing of financial instruments as a function of the risk factors. Chapter 5 proposes several methods used to approximate the portfolio distribution on the basis of both the distribution and the price information.

The approach used in *chapter 3, time series analysis*, assumes that the probability distribution of risk factors can be obtained from past observations of w. A time series model attempts to specify the *dynamics of the risk factors*, i.e. to specify a probabilistic model for the stochastic process which has generated the set of observations of interest. Since risk factor prices are quoted at discrete intervals of time, the main focus will be put on discrete time stochastic processes. It will be shown that the logarithmic price changes, the *log-returns*:

$$R(t, t+1) = \log\left(\frac{w(t+1)}{w(t)}\right) \qquad (1.14)$$

have more attractive statistical properties for the modelling than the prices themselves. The most convenient model for the log-returns is the assumption of joint normality. However, this assumption is not validated by empirical evidence which shows that the distribution of log-returns displays *leptokurtosis* (fat tails) and that squared log-returns are *autocorrelated*. The theoretic part of the chapter reviews the distributional characteristics of candidate models susceptible to capture the empirical features of financial return distributions. The empirical part presents an analysis of returns on Swiss stocks and compares the generic (normal) model with the alternative Student-t approach. Later on, the implications on risk management of the alternative modelling of the risk factors are discussed. The last section of the chapter gives insight into continuous-time modelling of stochastic processes which is of particular importance for the derivation of option pricing models.

Chapter 4 covers the valuation of financial instruments as functions of risk factors. Section 4.1 discusses the notion of *arbitrage*, which is crucial for the pricing of financial assets, in particular for derivative instruments. Generally speaking, arbitrage is defined as a trading strategy which generates a positive profit with no outlay of capital and no risk-taking. More specifically, Ingersoll (1987) identifies two kinds of arbitrage opportunities. Arbitrage opportunities of the first type are limited liability investments with no current commitment and a probability of profit. Arbitrage opportunities of the second type are limited liability investments with a current negative commitment and non-negative profits in the future. An economy whose agents are rational and non-satiated does not permit arbitrage opportunities. As a result, if no

arbitrage opportunities exist, restrictions can be placed on the asset price behaviour. In particular the price of any instrument which can be represented
as a combination of traded assets will be determined exactly. Moreover, the
lack of arbitrage opportunities ensures the existence of a *martingale* or *risk-
neutral* probability measure on Ω. Under the risk-neutral probability measure
asset prices are equal to the expected value of their pay-offs discounted at
the risk-free rate. The rest of the chapter is devoted to the pricing of specific
financial instruments. Note that there is a wide variety of financial assets
and the review will not be exhaustive. Most instruments, however, can be
decomposed into a small number of basic instruments, the *building blocks*,
which represent the main focus of this chapter. These building blocks are
identified as equities, bonds, forwards and options. The valuation of *equities*
is particularly simple to handle in the model since equity prices have been
modelled as risk factors. As a result, the value of any equity position reduces
to a multiple θ_i of the corresponding risk factor value w_i. A *bond* represents
an array of cash flows with different maturities. The valuation of bonds involves the discounting of each cash flow by an appropriate rate and requires
therefore the modelling of the term structure of interest rates.

Forward and futures contracts are agreements to buy a specified asset for a
given price on a specified date in the future. The derivation of forward prices is
easily obtained by the use of simple arbitrage arguments. Futures contracts
involve a daily settlement and have therefore a different cash flow profile
from this of a similar forward contract. However, it is generally assumed that
under suitable conditions the price difference between a futures contract and
a corresponding forward contract is small and is often ignored.

Options are contracts which give the right to buy or sell some underlying asset(s) for a specified price on a given future date. Option pricing is a
challenging task since it involves the modelling of the underlying price process. It will be shown that the price of an option is a non-linear function
of the underlying risk factor(s). Moreover, many option pricing equations
do not have a closed-form solution and have to be solved by the means of
numerical methods. For large portfolios these numerical evaluations require
a considerable computational effort. A widely used method which allows to
reduce the numerical burden consists in replacing the exact value function
(also called *full valuation*) by a global polynomial approximation based on a
Taylor expansion about the origin $w(0)$:

$$
g[w(T)] \approx g[w(0)] + \sum_{i=1}^{p} \frac{1}{i!} \left[\sum_{j_1=1}^{M} \sum_{j_2=1}^{M} \cdots \sum_{j_i=1}^{M} \frac{\partial^p g[w(0)]}{\partial w_{j_1} \partial w_{j_2} \cdots \partial w_{j_i}} d_{j_1} d_{j_2} \cdots d_{j_i} \right],
$$

(1.15)

where $d := w(T) - w(0)$.

Note that a polynomial approximation of order p requires the evaluation
of the portfolio and of its partial derivatives up to order p at the *supporting*

point $w(0)$ of the approximation. In practice, the approximation of order one, which will be called *linear approximation* in the following:

$$g[w(T)] \approx g[w(0)] + \left[\sum_{j_1=1}^{M} \frac{\partial g[w(0)]}{\partial w_{j_1}} d_{j_1} \right], \qquad (1.16)$$

and of order two, which will be referred to as *quadratic approximation*:

$$g[w(T)] \approx g[w(0)] + \left[\sum_{j_1=1}^{M} \frac{\partial g[w(0)]}{\partial w_{j_1}} d_{j_1} \right] + \frac{1}{2} \left[\sum_{j_1=1}^{M} \sum_{j_2=1}^{M} \frac{\partial^2 g[w(0)]}{\partial w_{j_1} \partial w_{j_2}} d_{j_1} d_{j_2} \right].$$
$$(1.17)$$

are of particular interest since the first and second partial derivatives, the "Greek letters" Delta and Gamma, are available for most option formulas. Polynomial approximations are typically accurate only in the vicinity of the supporting point. Section 4.5.2 discusses a possible improvement of the method by considering *piecewise approximations* of the pay-off profile, i.e. the use of several supporting points.

Chapter 5 reviews numerous methods used for the estimation of the portfolio distribution. An exact analytical representation of the portfolio distribution is not always possible since the pay-off profiles may be given only numerically. Moreover, under some models, which describe the dynamics of the risk factors, the unconditional distribution of the risk factors is not available. Even if an exact analytical representation of portfolio distribution would be possible, the approach would involve a considerable analytical effort. This effort is not justified if the portfolio of interest is often re-balanced, which leads to corresponding changes in the pay-off profile and in the risk factor sets.

Thus, one may at best obtain *analytical approximations* of the portfolio distribution by assuming jointly normal distributed risk factors and by replacing the pay-off profile by a global Taylor approximation. Analytical approximations based on a global linear approximation are referred to as *Delta approximations* and their analogues based on global quadratic approximation will be termed *Delta-Gamma approximations* .

Relaxing the assumptions concerning the risk factor distribution is achieved by considering *scenario-based methods*. The scenario-based methods require the generation of a set of market scenarios, i.e. of possible outcomes for the risk factor values. The scenarios are obtained through random sampling from the risk factor distribution which is the subject of section 5.2. The *Monte Carlo simulation* is a classical example of the scenario-based approach and will be discussed in section 5.3. The Monte Carlo simulation is based on a large set S with size N of scenarios for the risk factors and requires the full valuation of the portfolio for the N scenarios in S. The method may therefore involve a considerable numerical effort, especially for large portfolios with significant option content.

Section 5.4 presents a new scenario-based method called *Barycentric Discretisation with Piecewise Quadratic Approximation (BDPQA)* . The BDPQA involves the generation of two sets of scenarios S_1 and S_2. S_1 is a large set of size N^1, which corresponds to the set used in the Monte Carlo simulation. S_2 represents a small set with size J of distinguished scenarios for which the price function of the portfolio and its first and second derivatives are evaluated. The price function will then be approximated for each scenario in S_1 by the means of a quadratic approximation whose supporting point is a scenario in S_2. Thus, the BDPQA yields a piecewise quadratic approximation with J supporting points of the pay-off profile and requires therefore $J \ll N^1$ full valuations of the portfolio. The choice of the J supporting points, called *generalised barycenters*, is based on a subdivision of the risk factor space through simplicial partition.

Both, the BDPQA and the Monte Carlo Simulation, are based on the evaluation of a finite set of scenarios. As a result, their output yields a random sample whose *empirical distribution* gives an approximation of the true distribution. *Chapter 6* deals with the estimation of a risk measure (Value-at-Risk) from a sample. It analyses the effect of the sample size on the accuracy of the estimates and introduces sophisticated smoothing methods, based on weighted averages of *order statistics* or on *kernel smoothers* as possible enhancement for the risk measure estimation.

Finally, *chapter 7* summarises the results and gives an outlook to future work.

1.6 Test Environment

In order to compare alternative models of risk factor dynamics and to benchmark the different methods for evaluating F_g, two test environments have been developed. Both environments are pure option portfolios with highly non-linear risk profiles. In the first environment the risk factors are Swiss stocks, whose distributions are analysed in Chap. 3. The second environment is a FX-portfolio with 22 foreign currencies and the CHF as reference currency.

1.6.1 Environment I

The risk factors considered are the prices of 18 stocks of the Swiss Market Index. Four portfolios have been constructed, which consist in combinations of call and/or put options on the underlying stocks. Each option is evaluated with the Black-Scholes Formula and has a maturity of ten days.

Portfolio PF_1^I Portfolio PF_1^I consists of one at-the-money call option on each underlying stock. A long holding of the portfolio has an unlimited profit potential. The loss is limited by the amount paid for the options. The short holding has a profit limited to the premium received and an open-ended loss potential as the prices of the underlyings increase.

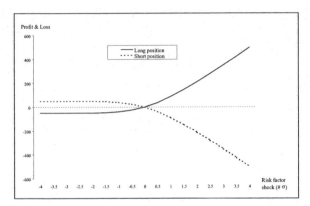

Figure 1.1. One-dimensional risk profile for portfolio PF_1^I

Portfolio PF_2^I Portfolio PF_2^I is a holding of one at-the-money put option on each risk factor. The long position increases in value as the market falls and the loss is limited by the initial value of the portfolio. The short position yields a maximum profit equal to the premium received and a nearly unlimited loss as the prices of the risk factors decrease.

Portfolio PF_3^I Portfolio PF_3^I is a combination of a long (short) call and a long (short) put with same strike price. This combination is referred to as a straddle. The long position has a loss potential bounded by the cost of the straddle and an open-ended profit potential. Such a strategy is appropriate when the market is expected to move significantly but the direction is not clear. The short position has obviously inverse profit and loss characteristics with unlimited loss potential and a maximal profit given by the profit received if the market stagnate.

Portfolio PF_4^I Portfolio PF_4^I is a inverse butterfly spread, which consists of four calls: one short (long) call with a low strike price A, two long (short) calls with middle strike price B and a short (long) call with high strike price

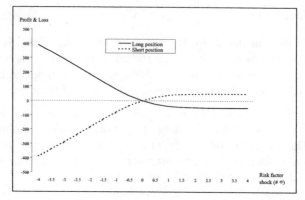

Figure 1.2. One-dimensional risk profile for portfolio PF_2^I

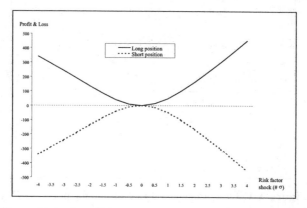

Figure 1.3. One-dimensional risk profile for portfolio PF_3^I

C. $B - A$ is equal to $C - B$. A long (short) position reaches a maximum loss (profit) if the market is at B at expiration. The loss (profit) is then equal to $B - A$ minus the price of the butterfly. The maximum profit (loss) occurring when the market moves significantly in either direction does not exceed the cost of the butterfly. Contrary to the straddle, the butterfly spread has limited potential for both profit and loss.

1.6.2 Environment II

Portfolio PF^{II} A FX-portfolio has been constructed which covers 23 exchange rates and contains 10'000 contracts on each exchange rate. The structure is the same for each currency. Half of the instruments are calls (puts). The calls and the puts have the same maturity structure: 40% of the instruments have a maturity of 7 days, 40% a maturity of 3 months, 10% a maturity of 6 months and 10% a maturity of one year. For each maturity, 40% of the derivatives are at-the-money, 20% are each mid in- or out-of-the

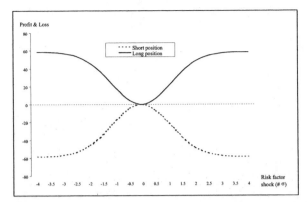

Figure 1.4. One-dimensional risk profile for portfolio PF_4^I

money and 10% are each deep in- or out-of-the money. As for each maturity and strike price the number of calls and puts are equal, the portfolio can be considered as a combination of straddles with limited loss potential and open-ended profit opportunity if the market moves significantly in either direction.

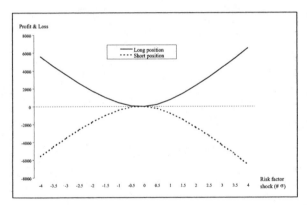

Figure 1.5. One-dimensional risk profile for portfolio PF^{II}

2. Risk and Risk Measures

As discussed in the introductory chapter, the uncertainty associated with the future (terminal) value of a portfolio may be modelled as a random variable $g[w(T)]$ whose probabilistic characteristics are specified by the induced probability measure P_g or its associated distribution function F_g. In some cases it might be more convenient to work with the portfolio profit and loss $\bar{g}[w(T)]$, respectively with the portfolio return $R_g[w(T)]$, and its induced probability measure $P_{\bar{g}}$, respectively P_{R_g}, or its associated distribution function $F_{\bar{g}}$, respectively F_{R_g}.

This chapter is devoted to the definition of *(market) risk*. Risk may be defined as the perception of uncertainty by individuals when they are faced with a decision problem. Two types of decisions, the *investment decision* and the *capital requirement decision*, which lead to different definitions of risk, will be discussed.

The *investment decision*, which is the topic of Sect. 2.1, represents the choice of a portfolio among an infinite set of possible combinations of financial assets. The investment decision requires the preference ordering of all alternative portfolios, the objects of choice, which are characterised by the probability measure or the probability distribution of their terminal value.

The most widely used modelling of individuals' preferences is based on the notion of *utility* and will be discussed in Sect. 2.1.1. The fundamental assumption underlying utility theory is that all interests of individuals for the objects of choice can be mapped into a single dimension called utility. Thus, the preference ordering of individuals can be represented by a utility function which assigns a value (a utility) to each alternative. When the alternatives are uncertain, such as in the investment decision, it will be shown that under certain assumptions concerning the (economic) human behaviour preferences can be represented by a specific type of utility function which is referred to as the *expected utility form*. Under the expected utility hypothesis the preference ordering is given by the expected value of a utility function u defined over the possible terminal values of the portfolios. The key assumptions underlying the expected utility hypothesis are however not uncontroversial. Section

2.1.1 discusses some critics which have been raised against these assumptions and reviews alternative models of economic human behaviour.

The topic of Sect. 2.1.1 is the attitude of individuals towards risk under the expected utility hypothesis. It will be shown that the curvature of the utility function, characterised by the second derivative of u, gives insight into the specific taste of individuals for risk. Section 2.1.2 discusses the ordering of portfolios under the expected utility hypothesis when only partial information is available about the utility function, such as the derivatives of u. The *stochastic dominance* rules, presented in Sect. 2.1.2, apply for the classes of utility functions consistent with the observed economic behaviour and poses no assumptions about the form of the portfolio distributions. The stochastic dominance rules involve the comparison of each pair of alternative portfolios. The *mean-risk analysis* discussed in Sect. 2.1.2 represents an alternative to the stochastic dominance approach by offering two-parameter rules for portfolio selection.

The second decision problem, the *capital requirement decision* discussed in Sect. 2.2, is conceptually less challenging than the investment decision since it does not involve a choice among alternatives and therefore does not require the modelling of preferences. The capital requirement decision is the assessment of the potential loss in value associated with a given portfolio at the terminal date T. In this approach risk is perceived as a kind of worst case loss and the corresponding risk measure estimates the appropriate amount of economic capital needed to cover the potential loss. Risk measures for capital requirement are therefore traditionally more conservative than their analogues for investment. Section 2.2.1 discusses the most widely used risk measure for capital requirement, Value-at-Risk (VaR), which is given with:

$$VaR_\alpha = v_\alpha = - \inf \{x \mid P\left(\bar{g}[w(T)] \leq x\right) \geq \alpha\}, \qquad (2.1)$$

Section 2.2.2 presents an axiomatic description of the properties that a meaningful measure of capital requirement has to possess. A risk measure satisfying the discussed axioms is called *coherent*.

2.1 The Investment Decision

Let's consider the general problem of the optimal allocation of scarce resources in a two period economy (see Ingersoll (1987); Merton (1992); Elton and Gruber (1995); Copeland and Weston (1992) for instance). Individuals in the first period invest (after consumption) their wealth $W(0)$ in assets and build a portfolio, which is sold at the beginning of the second period, at time T, to buy consumption goods. Two types of decisions have to be made. The first period decision is the choice of an optimal portfolio among the set of investment opportunities in order to transfer wealth from one period to the next. The second period decision is the allocation of the wealth among different consumption goods. Clearly, the investment decision is of interest here.

The investment decision represents the canonical problem of individual decision-making under uncertainty. The solution of the investment problem, i.e. the choice of the best investment mix requires both the definition of an *opportunity set* and of a *decision criterion*.

In general terms the *opportunity set*, also called *choice set*, represents the set of all alternatives which are open to individuals. In the investment decision the opportunity set G is the set of all feasible portfolios g^j, that is the set of those portfolios which may be constructed given the set of all available financial assets and given the *initial wealth* or *initial endowment* $W(0)$. It is assumed that the entire initial wealth is being invested and therefore the initial value of g^j, $g^j[w(0)]$, is equal to $W(0)$. Each *object of choice* g^j available to an individual is characterised by the induced probability measure P_{g^j} or the associated probability distribution F_{g^j} of its terminal value $g^j[w(T)]$. Thus, in this framework the agents choose the optimal portfolio by comparing the measures P_{g^j} or distributions F_{g^j}. It is further supposed that $g^j[w(t)]$ assumes value on some subset Z of $\mathbb{R}$. Hence, Z is the set of possible outcomes of the portfolios' terminal values which are also called *consequences* or *prizes* in literature. Let **P** represent the set of probability measures on Z. Thus, the set of probability measures induced by the feasible portfolios is a subset of **P**. In the theory of decision-making under uncertainty, the objects of choice are generally called *gambles, lotteries* or *(uncertain) prospects*. These three terms will be used interchangeably to design a feasible portfolio in the following.

The *decision criterion* represents the way individuals choose among the opportunity set G. It will be assumed that individuals' tastes may be summarised in a *preference relation* which in turn dictates the choice behaviour. A quantitative approach of the investment decision requires a numerical representation of the preferences, i.e. a mapping $U : G \to \mathbb{R}$ so that for $g^1, g^2 \in G$:

$$g^1 \text{ is preferred to } g^2 \Leftrightarrow U(g^1) > U(g^2) \qquad (2.2)$$

Section 2.1.1 discusses the axiomatic development, called expected utility hypothesis, which justifies the representation of preference ordering by U.

2.1.1 Utility Theory and Expected Utility Hypothesis

A commonly accepted theory of choice under uncertainty is the *expected utility hypothesis*. The expected utility hypothesis is a normative theory which is based on an axiomatic description of the human behaviour when faced with a choice. The analysis of the expected utility hypothesis will be split in two parts. Firstly, axioms allowing a formal description of the preference ordering in G will be introduced. Secondly, additional behavioural conditions, that are necessary and sufficient to obtain an expected utility representation, will be discussed.

Ordering Preferences and Utility Functions

A fundamental concept underlying the theory of preference ordering is that of *binary relation* on the opportunity set. Let's consider the Cartesian product $G \times G$ of G, i.e. the set of all ordered pairs (g^j, g^k) with $g^j, g^k \in G$. A *binary relation* on G is a subset of $G \times G$ (see Kreps (1988), p.7).

The expected utility hypothesis assumes that the preference ordering of individuals may be characterised by a binary relation $\succeq$ defined on the elements of G. Thus, $g^j \succeq g^k$ means that g^j is *(weakly) preferred* to g^k. Individuals are supposed to be *rational* in the sense that their preference ordering (or preference relation) exhibits following properties (see Huang and Litzenberger (1988), p.4):

- Axiom 1 *(Completeness)*:
  for any pair g^1, g^2: $g^1 \succeq g^2$ or $g^2 \succeq g^1$.
- Axiom 2 *(Transitivity)*:
  $g^1 \succeq g^2$ and $g^2 \succeq g^3$ imply $g^1 \succeq g^3$,

Hence, rationality is embodied in the assumptions of completeness and transitivity. The axiom of completeness postulates that individuals have well-defined preference between any two possible alternatives, i.e. that they are able to express preference over all pairs in $G \times G$. Transitivity suggests that individuals rank the alternatives in a consistent way. A cycling sequence of preferences of the form $A \succeq B, B \succeq C, C \succeq A$ is ruled out by the transitivity axiom.

Given preference relation $\succeq$, two other binary relations can be defined. g^1 and g^2 are said to be *indifferent* to each other if $g^1 \succeq g^2$ and $g^2 \succeq g^1$. The indifference relation is denoted by $g^1 \sim g^2$. g^1 is *strictly preferred* to g^2, noted $g^1 \succ g^2$, if $g^1 \succeq g^2$ but $g^2 \not\succeq g^1$ where $\not\succeq$ means not $\succeq$.

From the properties of the preference relation it can be easily shown that the strict preference relation is (see Kreps (1988), p.9):

1. asymmetric:
   $g^1 \succ g^2$ implies $g^2 \not\succ g^1$,
2. and negatively transitive:
   $g^1 \not\succ g^2$ and $g^2 \not\succ g^3$ imply $g^1 \not\succ g^3$.

Similarly, the indifference relation exhibits following properties:

1. reflexivity:
   $g^1 \sim g^1$,
2. symmetry:
   $g^1 \sim g^2$ imply $g^2 \sim g^1$,
3. and transitivity:
   $g^1 \sim g^2$ and $g^2 \sim g^3$ imply $g^1 \sim g^3$.

A mapping $U : G \to \mathbb{R}$ is referred to as a *utility function* representing the preference relation $\succeq$ if (see for instance Mas-Colell et al. (1995), p.9):

$$g^j \succeq g^k \Leftrightarrow U(g^j) \geq U(g^k) \qquad (2.3)$$

for all $g^j, g^k \in G$. Hence, U assigns a numerical value to each element of G and is therefore defined over the lotteries that is over the induced probability measures of the terminal values of the feasible portfolios. It may be shown (see Mas-Colell et al. (1995), p.9) that preference relation $\succeq$ can be represented by a utility function U only if $\succeq$ is complete and transitive. Note, however, that the assumptions of completeness and transitivity are not strong enough to ensure that any preference relation may be representable by a utility function. The additional assumption needed to ensure the existence of a utility function is called continuity and will be discussed in the next section. In case U exists it is *ordinal*, i.e. is unique up to a positive monotonic transformation. Thus, the numbers associated to the feasible portfolios allow only for a ranking. The magnitude of any difference in the ordinal utility measure is not meaningful. It is therefore impossible to deduce from the numbers assigned to the feasible portfolios how much more one is preferred to another.

Expected Utility Representation of Behaviour

As shown in the last section, the assumption of rationality allows the representation of preferences by a utility function U. However, U is defined over G and assumes therefore that individuals can express their preferences over probability distributions. This section discusses the additional assumptions which have to be made in order to represent the preference ordering of individuals as a function of monetary values, i.e. as a function of the possible terminal values of the feasible portfolios. It will be shown that the preference ordering may be summarised by the expected value of a utility function u defined over Z. The hypothesis of expected utility goes back to Bernoulli (1738). The axiomatic development of the expected utility hypothesis was first proposed by von Neumann and Morgenstern (1944), which imposed following axioms:

- Axioms 1-2 *(Rationality)*

  Individuals have a rational (complete and transitive) preference ordering $\succeq$ defined on G

- Axiom 3 *(Independence)*

  For all $p, q, r \in \mathbf{P}$ and $a \in (0, 1]$, $p \succ q$ implies $ap + (1 - a)r \succ aq + (1 - a)r$.

  The independence axiom means that if individuals prefer gamble p to gamble q, they will also prefer a gamble composed of $ap + (1 - a)r$ to a gamble composed of $aq + (1 - a)r$. The gambles $ap + (1 - a)r$ and $aq + (1 - a)r$ may be considered as compound lotteries (see Huang and Litzenberger (1988), p.8 for instance): a first lottery has two outcomes with probability a respectively $1 - a$ of occurrence. If the first outcome happens, lottery p (or q) is performed. If the second outcome happens, lottery r is performed. Note that r is independent of both p and q in the sense that the probabilities of occurrence of the outcomes of p and q are not conditioned on the probabilities of occurrence of the outcomes of r. The independence axiom suggests that lottery r is irrelevant for the comparison of lotteries p and q, that is independent lotteries do not intervene conjointly in the determination of preference.

- Axiom 4 *(Continuity)*

  For all $p, q, r \in \mathbf{P}$, if $p \succ q \succ r$ then there exists $a, b \in (0, 1)$ such that $ap + (1 - a)r \succ q \succ bp + (1 - b)r$.

  The continuity axiom postulates that no gamble p is so attractive that a compound lottery set up of p with (small) probability b and of r with probability $1 - b$ will always be preferred to q. Similarly, no lottery r is so unattractive that a compound lottery set up of p with probability a and of r with (small) probability $1 - a$ will never be preferred to q. The continuity axiom ensures the existence of a utility function representing $\succeq$.

Theorem 2.1 (von Neumann-Morgenstern) *A binary relation $\succeq$ on $\mathbf{P}$ satisfies Axioms 1-4, if and only if, there exists a function $u : Z \to R$ so that (see Huang and Litzenberger (1988), pp.7-8, Kreps (1988), p.46):*

$$p \succeq q \qquad iff \qquad \int_{z \in Z} u(z)dF_p(z) \geq \int_{z \in Z} u(z)dF_q(z) \qquad (2.4)$$

where $p, q \in \mathbf{P}$ and $\int_{z \in Z} u(z)dF_p(z) = U(p)$, $\int_{z \in Z} u(z)dF_q = U(q)$.

Note that the induced probability measure P_{g^j} of any feasible portfolio $g^j \in G$ belongs to $\mathbf{P}$.

Corollary 2.2 *Utility function $u(z)$ is unique up to a positive affine transformation. That is any function $v(z) = a + bu(z)$, where $a, b > 0$ are constant, represents the same preference ordering as $u(z)$.*

Corollary 2.3 *Individuals choose among the different portfolios so as to maximise the expected utility.*

A proof of theorem 2.1 can be found in von Neumann and Morgenstern (1947). Theorem 2.1 holds for arbitrary probability measures if Z has finite support. In case Z has infinite support, theorem 2.1 is applicable for the subset of **P** which contains the *simple probability* measures on Z. A probability measure $p \in \mathbf{P}$ is simple (see Kreps (1988), p.32 and appendix A.2) if $p(z) \neq 0$ for at most finitely many $z \in Z$ and $\sum_{z \in Z} p(z) = 1$. In both cases the expected utility hypothesis may be rewritten:

$$p \succeq q \text{ iff } \sum_{z \in Z} u(z)p(z) \geq \sum_{z \in Z} u(z)q(z). \tag{2.5}$$

The extension of theorem 2.1 for arbitrary probability measures when Z is infinite, which is rather technical, can be found in Kreps (1988) (pp.52-69) and in Fishburn (1970). The extension is generally based on an additional axiom called *dominance criterion*.

The expected utility hypothesis assumes that individuals express their preference over the set Z of terminal values of the feasible portfolios. Given the rational preference relation $\succeq$ and the axioms independence and continuity, the preference ordering can be represented by the expected value of the utility function u. u is often referred to as the *Bernoulli utility function* and U is termed *von Neumann-Morgenstern expected utility* function (see Mas-Colell et al. (1995), p.184 for instance). Corollary 2.2 implies that u, contrary to U, exhibits a cardinality property. The cardinality property allows not only the ranking of alternatives but also ensures that the magnitude of any difference in utility measure has a precise meaning. It should be noted that the expected utility hypothesis is not restrictive enough to specify a unique utility function but defines a whole class of admissible utility functions with various orderings of the feasible portfolios. Corollary 2.3 postulates simply that (rational) individuals always choose the most preferred alternative.

In the von Neumann-Morgenstern's approach uncertainty is *objective* in the sense that the probability of the events are given exogenously. Other approaches lead to an expected utility representation of the economic human behaviour. In particular the approach of Savage (see Savage (1954)) includes the *subjective* probability assessment as part of the investors' preference. However, the subjective probabilities must still satisfy the axioms of probability. Moreover, both taste, given by the utility measure, and beliefs, given by the probability measure, are independent (see Kreps (1988), p.34).

As a result, the distinction between subjective and objective probability assessment is inconsequent to the mathematical development of the expected utility representation of human behaviour.

Alternative Theories of Human Behaviour

Throughout the rest of this chapter it will be assumed that individuals are utility maximisers. However, it should be noted that the utility maximisation hypothesis is a normative theory based on the axioms discussed above. Many authors have cast doubts on the descriptive ability of the hypothesis. A classical case of violation of the expected utility axioms is the *Allais's paradox* (see for instance Spremann (1992)). The result of Allais (1953), which has been frequently replicated, shows that most individuals exhibit more preference for certain prospects than it would be expected by utility theory. An example of the Allais's paradox is given in Kreps (1988), pp.192-194. The behaviour described by the Allais's paradox is a violation of the substitution axiom.

Numerous other experiments have revealed deviations from the behaviour assumed by von Neumann-Morgenstern. The *behavioural finance* proposes alternative models of human behaviour. The models are often based on other fields of social sciences such as psychology. Some of the most important ones are discussed here briefly. The description is based on Schiller (1997), which offers an excellent review on the topic.

1. *Prospect Theory*

Prospect theory is the most widely used alternative to utility theory. It also gives a mathematical representation of human behaviour (see for instance Kahneman and Tversky (1979) and Tversky and Kahneman (1992)), as utility theory does. Under prospect theory individuals maximise a weighted sum of utilities, as it is the case under utility theory. The weights are given by a function of the "true" probabilities which gives no weight to events with extremely low probability of occurrence and a weight of one to events with very high probability of occurrence. The weighting function allows for explaining the Allais's paradox for instance. Moreover, under prospect theory the utility function, which is called the value function, is concave above a reference point and convex below it. The reference point is generally the current, or a psychologically important, wealth level. Thus, individuals are modelled as risk seekers below the reference point and as risk avoiders above it. This shape of the value function reflects loss aversion: where loss is involved individuals prefer a gamble with a significant probability of large loss but with a non-zero probability of gain to a smaller loss which occurs with certainty.

2. *Regret Theory*

Utility maximisers are immune to the pain of regret. Regret theory suggests that individuals are disappointed when they have made errors. For example they have a problem with the realisation of a loss and therefore defer selling assets that have gone down in value. They distinguish between "paper loss" and "realised loss". Loomes and Sugden (1982) propose a modified utility function which depends not only on a given choice but also on another choices under consideration.

3. *Mental Compartment*

One underlying assumption of expected utility theory is *asset integration*. Under this assumption individuals consider their investments as a unique portfolio and make choices by analysing the relevant aggregated outcomes. The hypothesis of mental compartment, also called mental accounting, suggests that people segment their investments into mental accounts. They make choices and evaluate gain and loss for each account separately. The hypothesis explains why numerous individuals segment their portfolio in a "safe" part with downside risk protection designed for saving and in a "risky" part designed for speculation. Similarly, risk managers frequently hedge specific transactions rather than considering the overall profit and loss situation of their portfolios.

Attitude Towards Risk

This section discusses the principal attitudes of expected utility maximisers towards risk. The theory of *risk attitudes* was developed by Arrow (1965) and Pratt (1964). Let's consider a twice differentiable (Bernoulli) utility function and a given level of wealth W. It is assumed that individuals are non-satiated, i.e. prefer more wealth than less. As a result, the first derivative of their utility function has to be positive: $u'(W) > 0$.

The specific investor's taste for risk places restrictions on the second derivative $u''(W)$ of the utility function. A fair gamble ε is defined as a random variable with $E[\varepsilon] = 0$ and $\text{var}[\varepsilon] > 0$. Clearly $E[W+\varepsilon] = W$. An investor is said to be *risk-averse* at a given level of wealth W if $u(W) > E[u(W + \varepsilon)]$ for all gambles. If the relation holds for each level of wealth, the investor is said to be globally risk-averse.

Hence, risk-averse individuals prefer receiving a fixed payment to a random payment of equal expected value. A sufficient and necessary condition for global risk aversion is a strictly concave utility function, that is $u''(W) < 0$ (which can be easily shown by using Jensen's inequality, see for instance Ingersoll (1987), p.37). *Risk-neutral* investors are indifferent to a fair gamble, which implies that $u''(W) = 0$. Investors with convex utility function, i.e. with $u''(W) > 0$, are referred to as *risk-seeking* investors.

Thus, the second derivative u'' of the utility function determines the taste for risk. It is generally assumed that individuals exhibit a risk-averse behaviour, i.e. have a strictly concave utility function. Recall that von Neumann-Morgenstern utility functions are only unique up to a positive affine transformation. u'' alone is therefore not sufficient to characterise the intensity of risk aversion. Let's consider $v(W) = a + bu(W), b > 0$. The behaviour associated with $v(W)$ is identical to the behaviour associated with $u(W)$. It is however obvious that:

$$v''(W) = bu''(W) \neq u''(W). \tag{2.6}$$

The intensity of risk aversion may be measured by the means of the function $A(W) = -u''(W)/u'(W)$ which is called the Arrow-Pratt *absolute risk aversion* (ARA) function (see Arrow (1965) and Pratt (1964)). Computing $A(W)$ for the utility function $v(W)$ yields:

$$A(W) = -\frac{bu''(W)}{bu'(W)} = -\frac{u''(W)}{u'(W)}. \tag{2.7}$$

Hence, $A(W)$ is free from scaling factors and therefore invariant to an affine transformation of the individual's utility function. The Arrow-Pratt measure may be used to evaluate the *risk premium*, i.e. the maximum amount that a risk-averse individual is willing to pay in order to avoid a gamble. Let's W_s be the level received under certainty so that:

$$E[u(W + \varepsilon)] = u(W_s). \tag{2.8}$$

W_s is referred to as the *certainty equivalent* of $W + \varepsilon$. Clearly, $\pi = W - W_s$ is the risk premium. Considering a Taylor expansion around W for the left-hand side of (2.8):

$$E[u(W + \varepsilon)] \approx E[u(W)] + u'(W)E[\varepsilon] + \frac{1}{2}u''(W)E(\varepsilon)^2$$
$$\approx u(W) + \frac{1}{2}u''(W) \text{ var } [\varepsilon], \tag{2.9}$$

respectively for the right-hand side of (2.8):

$$u(W_s) \approx u(W) - u'(W)\pi, \tag{2.10}$$

the risk premium can be approximated as follows:

$$\pi \approx \frac{1}{2} \text{ var } [\varepsilon] \left[-\frac{u''(W)}{u'(W)} \right]. \tag{2.11}$$

For non-satiated, risk-averse investors $A(W) = -u''(W)/u'(W) > 0$ and thus the risk premium is positive. Clearly, the value of the risk premium is dependent on the risk of the gamble, which is here given by the variance of ε and by the degree of risk aversion given by $A(W)$. The higher the absolute risk aversion, the higher the risk premium. It should be noted that for non-linear utility functions the Taylor expansion is accurate merely in the vicinity of W and therefore the approximation of the risk premium π given by the right-hand side of (2.9) performs well only for small risks.

However, the monotone relation between absolute risk aversion and risk premium holds globally. Actually, Pratt (1964) shows that given two monotone increasing and strictly concave utility functions u_1 and u_2 with corresponding ARA functions A_1, A_2 and risk premia π_1, π_2, following conditions are equivalent (for a discussion see Ingersoll (1987), p.40 and Takayama (1994), p.274):

1. $A_1(W) > A_2(W)$ for all W.
2. $\pi_1 > \pi_2$ for all W and for all gambles.
3. There exists an increasing, strictly concave function c, so that $u_1 = c(u_2)$.

The *relative risk aversion* (RRA) function $R(W)$, also introduced by Arrow and Pratt, is defined as follows (see for instance Ingersoll (1987), p.50):

$$R(W) = WA(W), \tag{2.12}$$

and $T(W)$ with:

$$T(W) = \frac{1}{A(W)}, \tag{2.13}$$

is referred to as the *absolute tolerance* function.

Utility functions with constant or decreasing ARA and constant or increasing RRA are consistent with the observed economic behaviour revealed by the risk attitude in Pratt (1964); Arrow (1965); Friend and Blume (1975) for instance. Note that a positive third derivative is a necessary condition for utility functions exhibiting non-increasing absolute risk aversion. For a given wealth level W:

$$
\begin{aligned}
A(W) &= -\frac{u''(W)}{u'(W)} \\
\frac{dA}{dW} &= -\left(\frac{u'''(W)u'(W) - [u''(W)]^2}{[u'(W)]^2} \right) \\
&= \left[\frac{u''(W)}{u'(W)} \right]^2 - \frac{u'''(W)}{u'(W)}.
\end{aligned}
\tag{2.14}
$$

Hence, since $u'(W)$ is positive, a positive $u'''(W)$ is necessary (but not sufficient) to have $\frac{dA}{dW} \leq 0$ and therefore a non-increasing absolute risk aversion.

A widely used class of von Neumann-Morgenstern utility functions is the *hyperbolic absolute risk aversion* (HARA) or *linear risk tolerance* (LRT) class of utility functions, which is defined as follows:

$$u(W) = \frac{1-\gamma}{\gamma} \left(\frac{aW}{1-\gamma} + b \right)^{\gamma} \qquad a, b > 0. \qquad (2.15)$$

Clearly:

$$u'(W) = a \left(\frac{aW}{1-\gamma} + b \right)^{\gamma-1},$$

$$u''(W) = -a^2 \left(\frac{aW}{1-\gamma} + b \right)^{\gamma-2},$$

$$T(W) = \frac{W}{1-\gamma} + \frac{b}{a},$$

$$A(W) = \frac{a}{\frac{aW}{(1-\gamma)} + b}. \qquad (2.16)$$

Hence, as the acronyms suggest, the absolute risk aversion, respectively the risk tolerance function is linear, respectively hyperbolic, in W. Special cases of HARA functions are listed in table 2.1.

Utility	U(W)	Parameter values	A(W)	R(W)
linear	a W	$\gamma = 1$	0 constant	0 constant
quadratic	$-\frac{1}{2}(-aW+b)^2$	$\gamma = 2$	$\frac{a}{b-aW}$ increasing	$W\left(\frac{a}{b-aW}\right)$ increasing
logarithmic	log(W)	$\gamma = b = 0$	$\frac{1}{W}$ decreasing	1 constant
power	$\frac{1-\gamma}{\gamma}\left(\frac{aW}{1-\gamma}\right)^{\gamma}$	$\gamma < 1$ b=0	$\frac{1-\gamma}{W}$ decreasing	$1-\gamma$ constant
negative exponential	$-e^{-aW}$	$\gamma \to -\infty$ b=1	a constant	aW increasing

Table 2.1. HARA utility functions

Obviously, the linear utility is the preference function of risk-neutral investors. The quadratic function plays a fundamental role in the mean-variance analysis. However, quadratic utility exhibits two undesirable properties. Firstly, the function is not increasing over the whole range of its domain ($W > \frac{b}{a}$) and is therefore not consistent with the non-satiation hypothesis.

Secondly, the function displays increasing absolute risk aversion. The last three functions are consistent with the economic behaviour observed in most empirical studies (ARA constant or decreasing, RRA constant or increasing).

Note that the use of single returns R_g with:

$$R_g = \frac{g[w(T)] - g[w(0)]}{g[w(0)]} = \frac{W(T) - W(0)}{W(0)}, \tag{2.17}$$

instead of the use of wealth levels is more convenient to compare portfolios since returns are scale-free. As the original $W(0)$ is known at time $t = 0$, the utility function may still be reexpressed as a function of the single returns. This convention will be followed in the rest of the chapter.

2.1.2 Rules for the Ordering of Uncertain Prospects

As shown in Sect. 2.1.1, investors whose behaviour obeys the axioms of von Neumann-Morgenstern choose the portfolio, from among the set of feasible portfolios, with the highest expected utility. The expected utility criterion allows for a complete ordering of the alternative portfolios. The (Bernoulli) utility function of individuals, however, is generally not known. At best only partial information such as the risk attitude of investors is available. This section discusses ordering rules which are equivalent to the expected utility criterion for specific classes of utility functions which are consistent with the observed economic behaviour discussed above. Following classes of utility functions are generally considered:

$$U_1 = \{u(x) \mid u'(x) > 0, \forall x \in \mathbb{R}\}$$
$$U_2 = \{u(x) \in U_1 \mid u''(x) < 0, \forall x \in \mathbb{R}\}$$
$$U_3 = \{u(x) \in U_2 \mid u'''(x) > 0, \forall x \in \mathbb{R}\} \tag{2.18}$$
$$U_4 = \left\{u(x) \in U_3 \mid -\frac{u''(x)}{u'(x)} \text{ decreasing}, \forall x \in \mathbb{R}\right\}$$

Clearly, U_1 represents the class of utility functions of non-satiated investors. U_2 is the subset of U_1 incorporating all (globally) risk-averse utility functions. U_3 includes all risk-averse utility functions with positive third derivatives. Finally, U_4 is a subset of U_3 with decreasing absolute risk-averse utility functions. As shown in (2.12) a positive third derivative is a necessary condition for utility functions displaying decreasing absolute risk aversion. Obviously, U_4 is the most restrictive class of utility functions but still remains consistent with the observed risk attitude discussed in Sect. 2.1.1.

Basically, two approaches allow ordering of the feasible portfolios defined above. The stochastic dominance rules are presented in Sect. 2.1.2 while the mean-risk criterion is discussed in Sect. 2.1.2. It is worth noting that, with only partial information that an individual's utility function belongs to a class defined above, it is not possible to achieve a complete ordering of the feasible portfolios. The ordering rules presented herein define a subset of the choice set G called the *admissible set* by discarding those feasible portfolios which are inferior (with respect to each utility function belonging to the class of interest) to a member of the choice set (see (Bawa (1975)).

Stochastic Dominance

The most general selection rule for ordering uncertain prospects in accordance with the von Neumann-Morgenstern framework is the stochastic dominance rule. Let's consider two uncertain prospects X and Y with distribution F and H defined over the range $[a, b]$. It will be assumed that distribution F (weakly) stochastically dominates distribution H to the nth order, if and only if, (see Ingersoll (1987), p.138):

$$
\begin{aligned}
F_{n-1}(x) &\leq H_{n-1}(x) && \text{for all } a \leq x \leq b, \\
F_k(b) &\leq H_k(b) && \text{for } k = 1, 2, \cdots, n-2,
\end{aligned}
\tag{2.19}
$$

where $F_0(x) = \int_a^x f(y)dy$ and $F_{n+1}(x) = \int_a^x F_n(y)dy$.

The stochastic dominance rule of order n provides a one-to-one correspondence with the expected utility rule for utility functions satisfying:

$$
(-1)^k u^{(k)}(x) \leq 0 \quad \text{for } k = 1, 2, \cdots, n,
\tag{2.20}
$$

where $u^{(k)}$ is the kth derivative of u, i.e. for utility functions whose derivatives alternate in sign.

Considering the classes of utility functions defined in the last section, the stochastic dominance rules can be summarised as follows:

1. *First Order Stochastic Dominance (FSD)*
   Probability distribution F is said to dominate probability distribution H in the sense of FSD, $F \underset{\text{FSD}}{\geq} H$, if and only if, $F(x) \leq H(x)$ for all values of x with strict equality for at least one value of x. FSD is the selection rule for the class of non-decreasing utility functions ($u' \geq 0$) and is therefore applicable for all utility functions in U_1. Thus, non-satiable individuals prefer F to H, if and only if, $F \underset{\text{FSD}}{\geq} H$. For proof see Hadar and Russel (1969); Hanoch and Levy (1969).

2. *Second Order Stochastic Dominance (SSD)*
   Probability distribution F is said to dominate probability distribution H in the sense of SSD, $F \underset{\text{SSD}}{\geq} H$, if and only if, $F_1(x) \leq H_1(x)$ for all values of x with strict equality for at least one value of x. SSD is the selection rule for the class of non-decreasing utility functions with non-positive second

derivative ($u' \geq 0, u'' \leq 0$) and is therefore applicable for all utility functions in U_2. Thus, non-satiable, risk-averse individuals prefer F to H, if and only if, $F \underset{\text{SSD}}{\geq} H$. For proof see Hadar and Russel (1969); Hanoch and Levy (1969).

3. *Third Order Stochastic Dominance (TSD)*
   Probability distribution F is said to dominate probability distribution H in the sense of TSD, $F \underset{\text{TSD}}{\geq} H$, if and only if, $F_2(x) \leq H_2(x)$ for all values of x with strict equality for at least one value of x and $F_1(b) \leq H_1(b)$, i.e. the mean of F is almost as large as the the mean of H. TSD is the selection rule for the class of non-decreasing utility functions with non-positive second derivative ($u' \geq 0, u'' \leq 0, u''' \leq 0$) and is therefore applicable for all utility functions in U_3. Thus, non-satiable, risk-averse individuals whose utility function has a positive third derivative, prefer F to H, if and only if, $F \underset{\text{TSD}}{\geq} H$.

   Bawa (1975) shows that third order stochastic dominance is a sufficient but not necessary rule for utility functions exhibiting decreasing absolute risk aversion ($u \in U_4$).

For the class of utility functions U_3, the third order stochastic dominance rules is a more useful selection criterion than the FSD and the SSD rules in the sense that the set of distribution functions which can be ordered by the TSD rule is greater than the set of distribution functions which can be ordered by the FSD or SSD rules. Thus, under the TSD rule the admissible set will be smaller than under the FSD or SSD rules.

The stochastic dominance approach is a very powerful tool which is equivalent to the expected utility criterion for arbitrary distributions provided that the utility function of interest belongs to one of the classes defined above. Since the SD criteria use the entire distribution they implicitly consider all moments. However, the approach has the major drawback of requiring pairwise comparisons of all alternatives. Since the opportunity set is infinite, the stochastic dominance rules are of little use for the investment decision problem.

Mean-Risk Analysis

Risk-averse, non-satiated investors will prefer return distribution F over H whenever (see Fishburn (1977)):

$$\mu(F) \geq \mu(H),$$
$$\text{and} \qquad (2.21)$$
$$r(F) \leq r(H),$$

where $\mu()$ is the mean of the distribution and $r()$ a risk measure defined as a function of the return distribution. Mean-risk analysis is closely related to the notion of *efficient portfolio*. Efficient portfolios are defined as portfolios for which there are no other portfolios with the same or greater expected return and less risk. The mean-risk criterion is not a complete ordering of all possible portfolio combinations but generates the admissible subset of portfolios which are efficient, the *efficient set* or *efficient frontier*. Rational investors choose one portfolio of the efficient frontier corresponding to their individual degree of risk aversion.

In the context of the mean-risk analysis the risk measure has often been characterised by the variance since the seminal work of Markowitz (1952). However, as discussed below, variance as a risk measure is only consistent with utility maximisation for a particular utility function or for a restricted class of return distributions. All risk measures discussed herein are special cases of Stone's generalised risk measure. Let's consider a portfolio with (single) return distribution F_{Rg} defined on the range $[a,b]$. The Stone's generalised risk measure on F_{Rg}, $r(F_{Rg})$, is given by (see Fishburn (1977)):

$$r(F_{Rg}) = \int_a^{\gamma(F_{Rg})} \mid \eta(F_{Rg}) - x \mid^\alpha dF_{Rg}(x) \qquad \alpha \geq 0, \qquad (2.22)$$

where $\eta(F_{Rg})$ is a reference level from which deviates are measured, $\gamma(F_{Rg})$ the range parameter specifying what deviations are to be included in the risk measure and α a measure of the relative impact of large and small deviations.

Note the generality of Stone's measure allows to choose the threshold $\eta(F_{Rg})$, from which an outcome is considered to be a risk, the part of the distribution which is considered in the risk measure with $\gamma(F_{Rg})$ and the relative weight α of extreme realisations with respect to the interior of the distribution.

Two widely used families of risk measures are derived from the generalised measure. The first one is the family of *central moments* of order n:

$$\mu_n = \int_a^b (x - \mu)^n dF_{Rg}(x), \qquad n \in \mathbb{N} \setminus \{1\} \qquad (2.23)$$

where $\gamma(F_{Rg}) = b$, $\eta(F_{Rg}) = \mu$ the mean of the distribution, and n an even integer. Note that for n odd $(x - \mu)^n = - \mid x - \mu \mid^n$ for all $x \leq \mu$ and $(x - \mu)^n = \mid x - \mu \mid^n$ for all $x \geq \mu$ and therefore the corresponding central moment is not a risk measure in the sense of Stone (see (2.22)). Obviously, central moments of odd orders are not adequate measures of risk since the positive and negative deviations from the mean have contributions of opposite sign.

The second special case of Stone's measure is the family of *lower partial moments* of order n defined as:

$$\text{LPM}_{n,l} = \int_a^l (l - x)^n dF_{Rg}(x).$$ (2.24)

As only the realisations of x which are smaller than l are considered, $l - x$ will be non-negative, so that $\text{LPM}_{n,l}$ is also for n odd a risk measure in the sense of Stone.

Central Moments: Variance and its Extensions The *variance*, the central moment of order two, is the traditional statistical measure of dispersion about the mean and a widely used measure of risk:

$$\text{var} = \int_a^b (x - \mu)^2 dF_{Rg}(x).$$ (2.25)

Modern Portfolio Theory, which has first established a formal risk-return framework bases since the work of Markowitz on variance as measure of risk. The most appealing feature of variance is its simplicity. Given a portfolio of N assets with a vector of portfolio weights θ and the variance covariance matrix Σ, the variance of the portfolio is:

$$\text{var} = \theta^\top \Sigma \theta = \sum_{i=1}^N \sum_{j=1}^N \theta_i \theta_j \sigma_{ij}.$$ (2.26)

Determining the efficient frontier reduces then to a quadratic programming problem. Variance as a criterion for risk, however, is only consistent with expected utility maximisation if restrictive assumptions are made about the utility function of the investor or about the distribution of the portfolio return. Let's consider following quadratic utility function at the return level R:

$$u(R) = R - \frac{cR^2}{2}.$$ (2.27)

The expected utility of a stochastic return $\tilde{R}$ with mean μ and variance σ^2 is therefore:

$$E\left[u(\tilde{R})\right] = \mu - \frac{c}{2}\left(\mu^2 + \sigma^2\right).$$ (2.28)

Thus, the expected utility depends only on the mean and variance of the distribution and, since c is positive for risk-averse utility functions, variance is the adequate measure of risk. For arbitrary distributions, the *quadratic utility* is the only utility function which justifies variance as a risk measure.

Quadratic utility, however, has serious drawbacks as the function is not increasing over the whole range of R and implies increasing absolute risk aversion. Nevertheless, numerous authors use the quadratic function as a Taylor approximation of order two of utility functions with more desirable properties. However, as shown in Brockett and Garven (1993), the approximation error due to the truncation of the Taylor expansion after two or even more terms can, for distributions and utilities commonly postulated in finance, become arbitrarily large.

The second justification for the use of variance as a risk measure is the assumption of normality of the portfolio returns. If the joint distribution of the returns of the risk factors is multivariate normal, the return of the portfolio containing instruments with linear pay-offs will be in turn normally distributed. As the normal distribution is completely described by its first and second moments, its variance is the adequate risk measure. In fact mean-variance analysis remains consistent with utility maximisation if the joint distribution of the risk factors belongs to the class of the *elliptically symmetric* or *elliptically contoured* distributions (see Ingersoll (1987), p.104). Members of this class have density function, if it exists, of the form (see Johnson (1987), p.107 and appendix A.3.2):

$$f(R) = k_M \, |\Sigma|^{-\frac{1}{2}} \, z \left[(R - \mu)^\top \Sigma^{-1} (R - \mu) \right], \qquad (2.29)$$

where z is a one-dimensional real-valued function k_M is a scalar proportionality constant which depends on the dimension M of the risk factor space, μ the mean vector and Σ is the non-singular dispersion matrix.

Elliptically symmetric distributions exhibit following properties (see Ingersoll (1987), pp.104-105 and Fang et al. (1990), pp.42-48):

- All marginal distributions from an elliptically symmetric distribution are elliptically symmetric with the same functional form.
- A linear combination of variables from an elliptically symmetric distribution has elliptically symmetric distribution with the same functional form and its parameters depend uniquely on the mean vector and the dispersion matrix.
- The dispersion matrix is always proportional to the covariance matrix when the variances exist and are finite.

Hence, in case the joint distribution of the risk factor returns is elliptically symmetric, the distribution of the portfolio return is one-dimensional elliptically symmetric and therefore completely described by its mean and its dispersion. Thus, for elliptically symmetric distributions with finite variance the mean-variance analysis remains appropriate.

Some special cases of the elliptically symmetric distribution are the multivariate normal, the multivariate Student-t and the multivariate Cauchy distributions. As discussed in the next chapter, symmetric distributions, however, are not adequate to describe the distribution of the risk factors. Asset prices are actually unbounded above but can not assume negative value. This implies a positive skewness for the price and the simple return distributions. The generic model of asset returns assumes for instance a lognormal distribution of the risk factor prices which in turn implies an approximative lognormal distribution of the portfolio containing instruments with linear pay-offs. Moreover, instruments with non-linear pay-offs such as options enable investors to create almost any arbitrary portfolio return distribution, which can definitely not be fully characterized by the first two central moments and for which the mean-variance criterion is no longer appropriate.

As the return distribution is generally non-symmetric, the third and higher central moments have been considered as potential risk measures. In particular *skewness* has been used as a measure of the asymmetry of a distribution. It should be stressed that the third central moment alone is not a measure of risk in the sense of Stone (see (2.22)) and skewness has therefore been considered as an extension of the traditional mean-variance analysis. The mean-variance-skewness analysis is consistent with expected utility maximisation for a broader class of return distributions than the mean-variance analysis. However, the three moment analysis is valid for arbitrary distributions only if investors have cubic utility functions (see Richter (1960); Levy (1969)):

$$u(R) = aR + bR^2 + cR^3.$$
(2.30)

By expressing $E[u(R)]$ in terms of the central moments $\mu_2 = E[R^2] - (E[R])^2$ and $\mu_3 = E[R^3] - 3\mu E[R^2]^2 + 2\mu^3$, it can easily be shown that the proper measure of risk is a combination of variance and skewness. Cubic utility functions, however, have two drawbacks: they do not exhibit decreasing marginal utility for all return levels and imply increasing absolute risk aversion within parts of their domain of definition. Moreover, the inclusion of skewness in the mean-risk analysis requires not only estimates of the single asset skewness but also of the joint movements (co-skewness). For a discussion see Ingersoll (1987), pp.99-101.

Lower Partial Moments An obvious drawback of variance and of any symmetric risk measure is that positive and negative deviations from the mean (or from any reference level) contribute equally to risk. Investors are naturally more concerned with the downside than the upside. Symmetric risk measures become unappropriate if used with non-symmetric distributions. The semi-variance criterion, which considers only negative deviations from the mean, has already be mentioned by Markowitz in its original work. More generally, risk measures considering the probability-weighted deviations below an arbitrary reference level or target l are referred to as Lower Partial Moments and have the form:

$$\text{LPM}_{n,l} = \int_a^l (l - x)^n dF_{R_g}(x), \qquad\qquad n \in \mathbb{N}_0. \qquad (2.31)$$

As shown above, all lower partial moments are risk measures in the sense of Stone (2.22) and coincide with the intuition which associates risk with failure to attain a target return (see Fishburn (1977)). Clearly, $\text{LPM}_{0,l}$ is the probability of the return falling below the target and is referred to as *probability of shortfall*. $\text{LPM}_{1,l}$ is the *target shortfall* and $\text{LPM}_{2,l}$ is the *lower partial variance* or *target semivariance*. The semivariance is a special case of the lower partial variance with $l = \mu$. Bawa (1975) shows that the set of utility functions for which the mean-lower partial moment rule is justified is extremely general. Considering classes of utility functions U_1, U_2 and U_3 defined in Sect. 2.1.2:

$$U_1 = \{u(x) \mid u'(x) > 0, \forall y \in \mathbb{R}\}$$
$$U_2 = \{u(x) \in U_1 \mid u''(x) < 0, \forall y \in \mathbb{R}\} \qquad (2.32)$$
$$U_3 = \{u(x) \in U_2 \mid u'''(x) > 0, \forall y \in \mathbb{R}\}$$

the dominance rules are summarised as follows:

- F is preferred to H for all utility functions in $U_i, i = 1, 2$, if and only if,

$$\text{LPM}_{i-1,l}(F) \leq \text{LPM}_{i-1,l}(H) \; \forall l \in \mathbb{R} \quad \text{and} \quad < \text{ for some } l.$$

- F is preferred to H for all utility functions in U_3, if and only if,

$$\mu_F \geq \mu_H$$

$$\text{LPM}_{2,l}(F) \leq \text{LPM}_{2,l}(H) \; \forall l \in \mathbb{R} \quad \text{and} \quad < \text{ for some } l.$$

Hence, $\text{LPM}_{0,l}$ is consistent with non-satiation, $\text{LPM}_{1,l}$ with risk aversion and $\text{LPM}_{2,l}$ with decreasing risk aversion. Moreover, the mean-lower partial moments analysis does not make any distributional assumption. However, to verify that F is preferred to H, the LPM functions must be computed and compared for every value of l. By evaluating the lower partial moment at a single l, only a subset of all admissible portfolios for the corresponding class of utility function can be obtained (see Bawa (1975)). Bawa (1978) shows that for arbitrary distributions this subset is a suitable approximation of the admissible set. For distributions completely characterised by a location and a scale parameter (and the distributions which belong to the location-scale family after a monotonic transformation of the random variables) the evaluation of the lower partial moments at a single point l is sufficient to obtain the set of all admissible portfolios (see Bawa and Lindenberg (1977)). Moreover, under appropriate distributional assumptions, the mean-LPM partial moment analysis includes the mean-variance analysis as a special case: Bawa and Lindenberg (1977) show that the admissible set under the mean-LPM_2 framework is the same as the admissible set under the mean-variance (or mean-location) framework if the distribution of the returns is normal or Student-t.

Another formal connection between mean-lower partial moments, stochastic dominance rules and utility theory can be found in Fishburn (1977). Proofs are provided that second order stochastic dominance implies LPM dominance for all degrees $n \geq 1$ and third order stochastic dominance for all degrees $n \geq 2$. It is further shown that the mean-LPM$_{n,l}$ criterion for an arbitrary l is equivalent to expected utility maximisation with utility functions of the form:

$$u(R) = R \qquad\qquad \text{for all } R \geq l \qquad (2.33)$$
$$u(R) = R - k(l - R)^n \qquad\qquad \text{for all } R < l \qquad (2.34)$$

for some positive constant k. Clearly, downside-risk avoiders have a risk-averse behaviour for $R < l$ and are risk neutral for $R \geq l$.

2.2 The Capital Requirement Decision

Risk measures for capital requirement are principally designed to determine the amount of capital needed to insure the institution's solvency, i.e. to cover potential losses. The capital requirement can be internal or regulatory. Internal capital requirement is the level of capital which the management consider to be appropriate. There is generally a trade-off between security and return on equity. A high amount of capital allows to absorb great losses but requires to earn high profit in order to maintain a target return on equity. Regulatory capital is imposed by some authority and forces financial institutions to maintain a certain level of capital. The main goal of regulators is the stability and integrity of the financial system. A major concern is the *systemic risk*, the risk that the default of one institution severely impairs the financial system. The deregulation and globalisation processes have increased the need for a supranational authority which is able to set unified standards for the participants of the financial market. Unified standards have the further advantage that competition is not being destroyed: an institution with lower regulatory capital requirement would be able to adopt a more aggressive pricing of its products.

Measures of capital requirement are also used to determine internal capital allocation and set position limits in the trading books of financial institutions. The use of a measure of potential loss to attribute capital or set limits is more meaningful than the traditional methods based on the size of the balance sheet, respectively on notional amounts.

Hence, risk measures for capital requirement evaluate the potential maximum loss of a portfolio value resulting from changes in the risk factor values at the end of a given holding period. The potential maximum loss at time T is given with:

$$\text{potential maximum loss} = \max\{-\bar{g}[w(T)] \mid \omega \in \Omega\}. \qquad (2.35)$$

The potential maximum loss may not be a very informative risk measure since the prices of financial instruments are not bounded above and it is therefore easy to construct portfolios (with short positions) for which an infinite loss has a non-zero probability of occurrence. To overcome this problem, two ways have been proposed. The first one considers a percentage point, a quantile of order α, of the profit and loss distribution. α is fixed somewhat arbitrary and varies generally from 1% to 5%. The method is referred to as Value-at-Risk. The second approach, the coherent risk measure approach, considers the supremum of the expected value of the loss over a non-empty set of probability measures defined on Ω. The approach is the result of an axiomatic definition of a meaningful (coherent) risk measure for capital requirement.

2.2.1 Value-at-Risk

Formally Value-at-Risk is given with:

$$\text{VaR}_\alpha = v_\alpha = -\inf\left\{x \mid P\left(\bar{g}[w(T)] \le x\right) \ge \alpha\right\}, \qquad (2.36)$$

or alternatively formulated:

$$\text{VaR}_\alpha = v_\alpha = -\inf\left\{x \mid F_{\bar{g}}(x) \ge \alpha\right\}. \qquad (2.37)$$

Value-at-Risk (VaR) represents a loss limit which is exceeded with respect to a (small) given frequency α. VaR is often defined as a measure of loss arising from *adverse market conditions* and is therefore related to the deviations from the nominal scenario $g[w^0] = g[E[w(T)]]$. In this case VaR yields:

$$\text{VaR'}_\alpha = v'_\alpha = -\inf\left\{x \mid P\left(g[w(T)] - g[w^0] \le x\right) \ge \alpha\right\}. \qquad (2.38)$$

The use of (2.36) or (2.38) as a definition of Value-at-Risk is inconsequential to the modelling and estimation methods presented in this book. VaR will refer to (2.36) in the sequel.

VaR is widely considered as the capital requirement method since it was proposed by the Basle Committee on Banking Supervision as an adequate risk measure for regulatory capital requirement (see BIS (1995c,b,a, 1996)). The quantitative criteria of the Basle Committee require that VaR is computed on a daily basis, using a level of $\alpha = 0.01$ for an horizon of at least 10 days. The capital charge for a bank is defined as the higher of:

- the previous day's Value-at-Risk;

- the average of the daily Value-at-Risk of the preceding sixty business days times a multiplicative factor.

The Basle Committee has concluded that a multiplicative factor of three (called hysteria factor by practitioners...) leads to a capital charge that provides a sufficient cushion for potential losses. Needless to say, a multiplicative factor of three provides an incentive to find aggressive VaR estimates.

Note that the Value-at-Risk approach says nothing about the potential size of the losses exceeding v_α. Moreover, VaR is in the center of a large debate since it fails to satisfy the subadditivity condition which will be formally defined in the next section. Briefly, the subadditivity property of a risk measure means that the evaluated risk of a combination of several portfolios should not be greater than the sum of the evaluated risks of the individual portfolios. The subadditivity condition is a natural requirement for the issues of capital allocation and position limits. Consider for instance a large financial institution with delocalised trading subsidiaries. Subadditivity means that the sum of the evaluated risks of the single subsidiaries gives an upper bound for the risk of the whole institution. Thus, it enables the delegation of risk measurement to the subsidiaries without loosing the oversight. As shown in Embrechts et al. (1998), VaR satisfies the subadditivity condition, and is coherent to be defined, if the risk factors are elliptically symmetric distributed.

In the event of $F_{\bar{g}}$ being strictly increasing, i.e. $F_{\bar{g}}^{-1}$ exists, it is easily verified from (2.37) that:

$$F_{\bar{g}}(-\text{VaR}_\alpha) = \alpha. \tag{2.39}$$

From (2.31), $\text{LPM}_{0,l}$, the probability of shortfall, applied to the profit and loss distribution is given with:

$$\text{LPM}_{0,l} = \int_a^l dF_{\bar{g}}(x). \tag{2.40}$$

By setting $l = -\text{VaR}_\alpha$, (2.40) becomes:

$$
\begin{aligned}
\text{LPM}_{0,-\text{VaR}_\alpha} &= \int_a^{-\text{VaR}_\alpha} dF_{\bar{g}}(x) \\
&= F_{\bar{g}}(-\text{VaR}_\alpha).
\end{aligned}
\tag{2.41}
$$

Hence, the relation between VaR and $\text{LPM}_{0,-\text{VaR}_\alpha}$ is immediate:

$$-\text{VaR}_\alpha = F_{\bar{g}}^{-1}(\text{LPM}_{0,-\text{VaR}_\alpha}). \tag{2.42}$$

2.2.2 Coherent Risk Measures

The axiomatic approach, proposed in Artzner et al. (1998), defines a set of properties that a meaningful risk measure for capital requirement, called coherent risk measure, should possess. A risk measure $r()$ is coherent (with respect of the profit and loss $\bar{g}[w(T)]$) if:

- Axiom 1 *(Translation invariance)*

  $r\left(\bar{g}[w(T)] + \alpha\right) = r\left(\bar{g}[w(T)]\right) - \alpha$, for all real number α.

  Translation invariance ensures that the addition of a sure amount α in the portfolio simply decreases the risk by α.

- Axiom 2 *(Subadditivity)*

  $r\left(\bar{g}_1[w(T)] + \bar{g}_2[w(T)]\right) \leq r\left(\bar{g}_1[w(T)]\right) + r\left(\bar{g}_2[w(T)]\right)$.

  As stated above, the subadditivity property ensures that the evaluated risk of a combination of several portfolios is not greater than the sum of the evaluated risks of the individual portfolios. Note that subadditivity is not only important for internal capital attribution but also for regulatory capital requirement. If the risk measure, used by a regulator to define capital requirement, does not possess the subadditivity property, the regulated institutions will have an incentive to set up subsidiaries in order to reduce their capital charges.

- Axiom 3 *(Positive homogeneity)*

  $r\left(\lambda\bar{g}[w(T)]\right) = \lambda r\left(\bar{g}[w(T)]\right)$, for all positive λ.

  Positive homogeneity requires the position size not to influence the risk measure directly. Clearly, this requirement assumes liquid markets.

- Axiom 4 *(Monotonicity)*

  $\bar{g}_1[w(T)] \leq \bar{g}_2[w(T)] \Rightarrow r\left(\bar{g}_2[w(T)]\right) \leq r\left(\bar{g}_1[w(T)]\right)$.

  The monotonicity axiom postulates that a risk measure should preserve the dominance ordering.

It is further shown in Artzner et al. (1998) that any coherent risk measure has the form:

$$r\left(\bar{g}[w(T)]\right) = \sup\left\{E_Q(-\bar{g}[w(T)]) \mid Q \in \mathcal{Q}\right\}. \tag{2.43}$$

where Q is a family of probability measures Q on Ω. A coherent risk measure is therefore the supremum of the expected value of the loss for a family of probability measures also called "generalised scenarios" on Ω. A generalised scenario can be for example a scenario in the conventional sense, a realisation of the risk factors for which the portfolio is evaluated. In this case the corresponding probability measure is a point mass. The coherent risk measure is the worst case scenario of the set considered. Clearly, the more scenarios are considered, the more conservative the risk measure is. The major drawback of conventional scenario analysis is the rather arbitrary choice of the set of scenarios. Nothing is said about the likelihood of the different scenarios. Hence, the distribution information is lost.

However, since the real measure P is in Q, the expected shortfall:

$$r\left(\bar{g}[w(T)]\right) = -E\left(\bar{g}[w(T)] \mid \bar{g}[w(T)] \leq l\right), \qquad l < 0 \qquad (2.44)$$

is coherent. Artzner et al. (1998) propose to set $l = -\text{VaR}_\alpha$ and define so the tail conditional expectation, TailVaR_α or tv_α for short, by:

$$\text{TailVaR}_\alpha = -E\left(\bar{g}[w(T)] \mid \bar{g}[w(T)] \leq -\text{VaR}_\alpha\right). \qquad (2.45)$$

The tail conditional expectation is therefore a coherent extension of VaR, which measures the expected loss, given the loss exceeds VaR. It should be considered that the estimation of an expectation beyond VaR requires a very careful modelling of the left tail of the profit and loss distribution.

Note the relation between TailVaR, VaR, and the first and second lower partial moments of the profit and loss distribution:

$$\begin{aligned}
\text{LPM}_{1,l} &= \int_a^l (l - x)\, dF_{\bar{g}}\left(x\right) \\
&= l\int_a^l dF_{\bar{g}}\left(x\right) - \int_a^l x\, dF_{\bar{g}}\left(x\right) \\
&= lP\left(\bar{g}[w(T)] \leq l\right) - E\left(\bar{g}[w(T)] \mid \bar{g}[w(T)] \leq l\right) P\left(\bar{g}[w(T)] \leq l\right) \\
&= \left(l - E\left(\bar{g}[w(T)] \mid \bar{g}[w(T)] \leq l\right)\right) P\left(\bar{g}[w(T)] \leq l\right) \\
&= (l + \text{ expected shortfall})\, \text{LPM}_{0,l}.
\end{aligned}$$
$$(2.46)$$

For $l = -\text{VaR}_\alpha$, (2.46) becomes:

$$\text{LPM}_{1,-\text{VaR}_\alpha} = (-\text{VaR}_\alpha + \text{TailVaR}_\alpha)\text{LPM}_{0,-\text{VaR}_\alpha}, \qquad (2.47)$$

and hence:

$$\text{TailVaR}_\alpha - \text{VaR}_\alpha = \frac{\text{LPM}_{1,-\text{VaR}_\alpha}}{\text{LPM}_{0,-\text{VaR}_\alpha}}. \qquad (2.48)$$

An alternative risk measure for capital requirement that satisfies the sub-additivity condition is *Maximum Loss* (ML), which represents the potential maximum loss over a subset of the risk factor space. Formally, $\bar{\Omega}$ is defined as a compact subset of Ω, so that $P\left(\omega \in \bar{\Omega}\right) = 1 - \alpha$ with $0 < \alpha < 1$. Maximum Loss is defined by:

$$\text{ML}_\alpha = \max\{-\bar{g}[w(T)] \mid \omega \in \bar{\Omega}\}. \qquad (2.49)$$

$\bar{\Omega}$ is referred to as the trust region and Maximum Loss represents the loss for the worst case scenario on $\bar{\Omega}$. Note that the trust region is not uniquely defined. In (the very simple) case the risk factor space is characterised by a one-dimensional standard normal distribution, it is obvious that for instance the three following ranges $[-\infty, 1.65]$, $[-1.96, +1.96]$ and $[-1.65, +\infty]$ define a 95% trust region of the risk factor space. For symmetric elliptical distributions (see Sects. 2.1.2 and A.3.2), a possible choice for the trust region is the region delimited by the ellipsoids that represent the contours of constant density. As already mentioned in Sect. 2.1.2 and as discussed in the next chapter, symmetric distributions, however, are not adequate to describe the dynamics of the risk factors.

Remark that:

$$P\left(-\bar{g}[w(T)] \leq \text{ML}_\alpha\right) = P\left(\omega \in \bar{\Omega}\right) + P\left(-\bar{g}[w(T)] \leq \text{ML}_\alpha \mid \omega \in \bar{\Omega}^c\right)$$
$$\geq 1 - \alpha, \qquad (2.50)$$

and therefore:

$$P\left(-\bar{g}[w(T)] > \text{ML}_\alpha\right) \leq \alpha. \qquad (2.51)$$

From the definition of Value-at-Risk (see (2.36)) and for a strictly increasing distribution function:

$$P\left(-\bar{g}[w(T)] > \text{VaR}_\alpha\right) = \alpha, \qquad (2.52)$$

it results that $\text{ML}_\alpha \geq \text{VaR}_\alpha$. Hence, Maximum Loss is a more conservative risk measure than Value-at-Risk. Note that ML gives no insight into the size of potential loss, provided that $\omega \in \bar{\Omega}^c$.

The evaluation of Maximum Loss requires the maximisation of a non-convex high dimensional value function. Solutions for the case of quadratic risk profiles and joint normally distributed risk factors can be found in Studer (1997).

2.3 Summary

The principal objective of the investment decision is the choice of the most preferred portfolio from among the infinite set G. In an economy where individuals obey the axioms of von Neumann-Morgenstern, preferences have an expected utility representation. The decision criterion is by nature subjective since utility is specific to individuals. For restricted classes of utility functions that are consistent with the observed economic behaviour the stochastic dominance rules are equivalent to the expected utility criterion and applicable to arbitrary distributions. They require, however, pairwise comparisons of the alternatives and are not feasible for the whole opportunity set. The use of the stochastic dominance approach is therefore restricted to the analysis of those small subsets of G which are of particular interest. The mean-risk analysis reduces the investment decision to a two-dimensional problem. In this approach risk is explicitly defined as a probability weighted deviation from an arbitrary reference level. The mean-variance criterion, which is the most widely used ordering rule in finance, is motivated by assuming either symmetry of the portfolio distributions or quadraticity of the individuals' utility functions. These strong assumptions can be relaxed by considering lower partial moments as the risk measure. By evaluating the LPM at a single reference level, the mean-lower partial moment criteria represent a suitable approximation of the stochastic dominance rules.

The risk measures for the capital requirement decision assess the potential loss on a given portfolio at some terminal date T. Value-at-Risk is the most widely used risk measure in this framework and has become the finance industry standard. Value-at-Risk, however, has theoretical drawbacks such as the lack of the subadditivity property. TailVaR represents an attractive alternative to VaR which gives insight into the expected loss, given the loss exceeds VaR.

Most risk measures presented in this chapter focus on the downside portion of the portfolio (respectively return or profit and loss) distribution and therefore require the careful modelling of the left tail of the distribution. If no assumption can be made about the distributional form of the portfolio, the evaluation of risk requires therefore a full description of the distribution of interest. In other words, a partial characterisation of uncertainty by the means of a finite set of central moments will not be sufficient for an adequate assessment of risk.

3. Modelling the Dynamics of the Risk Factors

The purpose of this chapter is to characterise the source of uncertainty in our model, the value $w(T)$ of the risk factors at some future date T. Current prices are generally supposed to reflect all available relevant information, which arrives in a random fashion. As a result, future prices seem to be very unpredictable. The empirical evidence shows that attempts to forecast precise values, *point estimates*, for $w(T)$ are not very successful. A better approach consists in describing $w(T)$ by a probability distribution. This in turn requires the building of an adequate model which describes the dynamics of the risk factors, i.e. their evolution through time. Basically two types of approaches can be considered for the building of a model. An *econometric model* attempts to formulate the behaviourial relationship between $w(T)$ and some other explanatory economic variables. A *time series model* examines the past behaviour of w in order to infer about its future. The latter approach is used in this chapter. More concretely, a time series model attempts to specify a probabilistic model for the dynamics of the risk factors, a *stochastic process*, that could have generated the set of observations considered.

The portfolio approach requires the knowledge of the M-dimensional joint distribution of the risk factors. However, in order to focus on the dynamic properties of the stochastic processes presented, the theoretic part of this chapter reviews one-dimensional (univariate) models. The multivariate extensions are provided for the models implemented in the empirical part. Since prices can be observed only in a discrete way the models presented herein are mostly discrete. The transition to continuous-time models is discussed at the end of the chapter.

Section 3.1 develops the basic concepts needed for the analysis of *univariate stochastic processes* in the *time domain* and introduces the conventional ARMA approach. The specific modelling of financial series is discussed in the latter part of the section, where it is suggested that the daily logarithmic price changes (log-returns) have more attractive statistical properties than the prices themselves. The formulation of an adequate model begins with some theoretical arguments about the price behaviour in financial markets. The *efficient market hypothesis*, which is subject of Sect. 3.2, asserts that past returns can not be used to improve the prediction of future returns. Based on the efficient market hypothesis, two theoretical models for the evolution of the logarithmic price changes have been proposed in the literature. The *random walk* model assumes that the log-returns are independent of each other. The less restrictive *martingale hypothesis* rules out the presence of serial correlation in a log-returns series. The lack of serial correlation has been largely confirmed by empirical evidence. Section 3.3 reviews other *empirical features* of financial return distributions: calendar effects, leptokurtosis of the unconditional distributions and serial correlation of the squared log-returns. Section 3.4 proposes candidate models for returns which are consistent with some of the major features identified in Sects. 3.2 and 3.3. It will be shown that models allowing for changing the conditional volatility can accommodate excess kurtosis and serial correlation of the squared returns. Section 3.5 presents an empirical analysis of the returns on Swiss stocks and compares the generic (Gaussian) model with the Student-t approach. The implications for risk management of the alternative modelling of the risk factors are discussed at the end of the section. Section 3.6 gives insight in the continuous-time modelling of stochastic processes by analysing the properties of the Brownian motion, the continuous-time equivalent of a random walk. The continuous-time modelling is of particular importance for the derivation of option pricing models, which will be discussed in the next chapter.

This chapter has to be read in the light of risk management, the following characteristics of which are worth being recalled:

- As shown in the latter chapter, several risk measures like VaR, TailVaR and LPM focus on the left tail of the portfolio distribution. As a result, particular care should be given in the tails of the risk factor distribution. This is of importance since, as it will be shown, risk factor distributions are fat-tailed.
- Risk management deals with very large portfolios of financial assets which depend on hundreds of risk factors. Moreover, the portfolios are permanently re-balanced, which induces permanent changes of the underlying risk factor set. The models used for the risk factor distribution should therefore be as simple as possible in order to allow frequent reevaluations of the risk exposure.

3.1 Statistical Definitions

3.1.1 Stochastic Processes: Basic Definitions

Let's consider the observed univariate series $\{x(1), x(2), \cdots, x(n)\}$ where $x(t)$ is a single number recorded at time t and the observations are available for n consecutive times (see Taylor (1986), p.16). The value of $x(t)$ is not known before time t. In time series analysis $x(t)$ is generally considered as a particular realisation of the stochastic variable $X(t)$. If observations for an infinite period of time were available (see Hamilton (1994), p.43):

$$\{x(t)_{t=-\infty}^{t=+\infty}\} = \{\cdots, x(-2), x(-1), x(0), x(1), x(2), \cdots, x(n), \cdots\}. \quad (3.1)$$

The infinite sequence $\{x(t)_{t=-\infty}^{t=+\infty}\}$ would be viewed as a single realisation of the stochastic process $\{X(t)_{t=-\infty}^{t=+\infty}\}$. Formally a stochastic process (see Roger (1991), p.167) is a collection of random variables $\{X(t), t \in \mathcal{T}\}$ defined on $(\Omega, \mathcal{F}, P)$ and assuming values in $(\Omega', \mathcal{B})$. Ω' is generally the real line for one-dimensional processes and $\mathbb{R}^M$ for M-dimensional processes. For time series $\mathcal{T}$ represents the time. The process is *continuous* if $\mathcal{T}$ is the real line and *discrete* if $\mathcal{T}$ is a subset of the integers. If the values of the process are countable, it is referred to as a *discrete-state* process. Otherwise it is a *continuous-state* process.

The aim of time series analysis is the description of the stochastic process that is assumed to be generating the observed series of interest. A fully characterisation of a stochastic process requires the specification of the joint distribution of the $X(t)$. Time series analysis focusses generally on a partial description of the process consisting in the first and second moments. These moments are defined with respect to the moments of the random variables $X(t)$ (see Granger and Newbold (1986), p.3).

The *mean* $\mu(t)$ of process at time t is:

$$\mu(t) = E[X(t)]. \quad (3.2)$$

The *covariance* $\lambda_{t,t-\tau}$ between $X(t)$ and the *lagged variable* $X(t-\tau)$ is given by:

$$\lambda_{t,t-\tau} = \text{cov}[X(t), X(t-\tau)] = E[(X(t) - \mu(t)(X(t-\tau) - \mu(t-\tau)], \quad (3.3)$$

so that $\lambda_{t,t}$ is the variance of the process. The covariance is often referred to as the *autocovariance* or *serial covariance*. It is generally impossible to obtain more than one realisation of the stochastic process, and this only for a finite time period. It is therefore impossible to make some inference about the moments of the process unless some restrictions like *stationarity* and *ergodicity* can be placed on the process. Stationarity implies that the distribution of $X(t)$ is the same for all t. Ergodicity means that the sample moments of a sequence of observations converge (in probability) to their population counterparts. If stationarity and ergodicity can be assumed the observations over time can therefore be considered as a sample from a given population. Conventional statistical methods designed to infer properties of the population from the sample are valid. Stationarity and ergodicity are discussed more formally in the next section.

3.1.2 Properties of Stochastic Processes

A stochastic process is said to be *strictly stationary* (see Taylor (1986), p.16) if for all integers i, j all integers k the multivariate distribution function of $(X(i), X(i+1), \cdots, X(i+k-1))$ is identical to that of $(X(j), X(j+1), \cdots, X(j+k-1))$. This means that shifting the time origin from i to j has no effect on the distribution, which depends only on the intervals $i, i+1, \cdots$ Note that the definition holds for $k = 1$ and stationarity implies that the distribution of $X(t)$ is the same for all t. Thus, stationarity requires that the process which generates the series is invariant with respect to time translation. Obviously, for a stationary process following conditions are satisfied:

$$E[\mu(t)] = \mu$$
$$\text{cov}[X(t), X(t-\tau)] = E[(X(t) - \mu)(X(t-\tau) - \mu)] = \lambda(\tau). \tag{3.4}$$

Thus, the first two moments, provided they are finite, are constant and invariant of t. The autocovariances $\lambda(\tau)$ depend only on τ, the *time lag*. It is easily verified with (3.4) that $\lambda(\tau) = \lambda(-\tau)$. Strictly speaking, stochastic processes with fixed first and second moments are said to be *second order, covariance or weak stationary*. If joint normality can be assumed, so that the distribution is entirely characterised by the two first moments, weak stationarity implies strict stationarity. As the variance of a weak stationarity process is constant and equal to $\lambda(0)$, the autocovariance at lag τ may be standardised by dividing it through the variance. This yields *autocorrelation* $\rho(\tau)$:

$$\rho_\tau = \frac{\text{cov}[X(t), X(t-\tau)]}{\sigma(t)\sigma(t-\tau)} = \frac{\lambda(\tau)}{\lambda(0)}, \tag{3.5}$$

with $-1 \leq \rho(\tau) \leq 1$. A process is called uncorrelated if the correlation between $X(t)$ and $X(t-\tau)$ is zero for all τ. The time properties of a stochastic process are generally summarised either by the autocovariance or by the autocorrelation function, i.e. by plotting $\lambda(\tau)$ respectively $\rho(\tau)$ against τ.

A covariance-stationary process is *ergodic for the mean* if (see Hamilton (1994), p.47) the sample mean of the observation converges in probability to $E[X(t)]$:

$$\bar{\mu} = \frac{1}{n} \sum_{t=1}^{n} X(t) \xrightarrow{P} E[X(t)]. \tag{3.6}$$

Additionally, a covariance-stationary process is *ergodic for the second moments* if

$$\bar{\lambda}(\tau) = \frac{1}{n-\tau} \sum_{t=\tau+1}^{n} X(t) \xrightarrow{P} \lambda(\tau), \tag{3.7}$$

for all τ. Hence, if the process is ergodic the sample estimates $\bar{\mu}$ and $\bar{\lambda}(\tau)$ are consistent estimates of the mean respectively of the autocovariances. Note that with a single realisation from the process it is not possible to test for ergodicity (see Granger and Newbold (1986), p.5).

The observations generated by a (weak) stationary process will fluctuate around the mean with no tendency for their dispersion to vary over time. The assumption of stationarity is however often too restrictive for financial time series. It may therefore be appropriate to transform the series by applying operations like differences or logarithms until it is approximately stationary. An adequate transformation for the risk factor series is discussed in Sect. 3.1.3. As usual in time series analysis it will be assumed that stationarity implies ergodicity.

It should be noted that stationarity implies that the unconditional distribution of the $X(t)$ remains constant. However, the distribution and hence the moments conditional to some information set I_t, may differ from the stationary distribution. One example for the conditional mean of an AR(1) process is given in (3.24) below. In time series analysis a typical information set I_t is the set of past realisations $x(s)$ with $s < t$, i.e. the natural filtration of $X(t)$ (see appendix A.2.5).

3.1.3 Basic Stochastic Processes

The fundamental *Wold's Decomposition theorem* states that any zero-mean covariance-stationary process $\{X(t)\}$ can be represented in the form (see Hamilton (1994), p.109):

$$X(t) = \sum_{j=0}^{\infty} \psi_j \varepsilon(t - j) + \kappa(t), \qquad (3.8)$$

where $\psi_0 = 1$ and $\sum_{j=0}^{\infty} \psi_j^2 < \infty$. The uncertainty derives from the variable $\varepsilon(t)$ which is an innovation with mean zero and variance σ^2 (see Sect. 3.1.3). The remaining term of the model is $\kappa(t)$ which is uncorrelated with $\varepsilon(t - j)$ for any j and can be predicted arbitrary well from a linear function of past values of X:

$$\kappa(t) = \hat{E}\left[\kappa(t) \mid X(t - 1), X(t - 2), \cdots\right]. \qquad (3.9)$$

$\kappa(t)$ is called the linearly deterministic component of $X(t)$.

White Noise Process A (zero-mean) *white noise process*, which is the simplest stationary process, is a sequence $\{\varepsilon(t)\}$ so that (see Hamilton (1994), p.47):

$$E[\varepsilon(t)] = 0,$$
$$E[\varepsilon(t)^2] = \sigma^2, \qquad\qquad (3.10)$$
$$E[\varepsilon(t)\varepsilon(\tau)] = 0 \qquad \text{for } t \neq \tau.$$

Hence a white noise process is a sequence of uncorrelated random variables $\varepsilon(t)$ with zero mean and constant variance σ^2. The $\varepsilon(t)$ are often referred to as *shocks, innovations or random disturbances*. Zero autocorrelation does not necessarily imply the independence of $\varepsilon(t)$ and $\varepsilon(\tau)$ for $t \neq \tau$. If, however, the joint distribution of $\varepsilon(t)$ is multivariate normal, the process is called Gaussian, and zero autocorrelation is equivalent to independence. A white noise process with independent $\varepsilon(t)$ is termed strict *white noise*. As (3.8) shows, the property of a stationary time series can be reproduced by introducing an infinite number of lagged innovations $\varepsilon(t - j)$ with $j > 0$. Thus, white noise processes $\{\varepsilon(t)\}$ are used as a basic tool to construct new processes. In the following the symbol $\varepsilon(t)$ will always denote an innovation with a mean zero and variance σ^2.

ARMA Processes

Moving-Average Process A process $\{X(t)\}$ defined by :

$$X(t) - \mu = \varepsilon(t) + \beta_1\varepsilon(t - 1) + \cdots + \beta_q\varepsilon(t - q) \qquad (3.11)$$

is called a *moving average process* of order q and denoted by $\mathrm{MA}(q)$. A $\mathrm{MA}(q)$ process is completely described by a weighted sum of lagged random disturbances. $\mathrm{MA}(q)$ models are therefore used for time series, which are affected by a variety of random events (new information) whose effect is not only immediate but may affect the time series in several subsequent periods as the market assimilates the relevance of the news.

$\mathrm{MA}(q)$ processes inherit the stationarity property of the white noise processes. The first moments of the process are (see Hamilton (1994), p.51):

$$E[x(t)] = \mu,$$
$$\lambda(0) = \left(1 + \sum_{i=1}^{q} \beta_i^2\right)\sigma^2,$$
$$\lambda(\tau) = \begin{cases} \left(\sum_{i=\tau}^{q} \beta_i\beta_{i-\tau}\right)\sigma^2 & \text{for } \tau \leq q \text{ and with } \beta_0 = 1, \\ 0 & \text{for } \tau > q. \end{cases} \qquad (3.12)$$

The $\mathrm{MA}(q)$ has therefore a memory of exactly q periods: the covariance is zero for every lag $\tau > q$.

Autoregressive Process A p-order *autoregressive process* AR(p) is characterised by :

$$X(t) - \mu = \sum_{i=1}^{p} \alpha_i(X(t-i) - \mu) + \varepsilon(t). \qquad (3.13)$$

Hence, the model depends on a weighted sum of past values, the lagged variables, and is adequate to many situations in which it is reasonable to assume that the present value of a time series depends on the immediate past values.

Certain restrictions must be placed on the parameters α_i to ensure stationarity. In mathematical terms (3.13) is a pth-order difference equation. The process turns out to be stationary if the difference equation is stable, i.e. if all of its solutions are bounded (see Kincaid and Cheney (1996), p.34). This is achieved whenever the roots of the characteristic equation:

$$1 - \alpha_1 z - \alpha_2 z^2 - \cdots - \alpha_p z^p = 0 \qquad (3.14)$$

lie outside the unit circle.

Considering the AR(1) process:

$$X(t) - \mu = \alpha_1(X(t-1) - \mu) + \varepsilon(t), \qquad (3.15)$$

the characteristic equation is obviously:

$$z = \frac{1}{\alpha_1}. \qquad (3.16)$$

Hence, stationarity requires that $|\alpha_1| < 1$. Substituting in (3.15) for lagged values of $X(t)$ gives:

$$
\begin{aligned}
X(t) - \mu &= \alpha_1(X(t-1) - \mu) + \varepsilon(t) \\
&= \alpha_1\left[\alpha_1(X(t-2) - \mu) + \varepsilon(t-1)\right] + \varepsilon(t) \\
&= \cdots \qquad\qquad\qquad\qquad\qquad\qquad\qquad\qquad (3.17) \\
&= \alpha_1^I X(t-I) + \sum_{i=0}^{I-1} \alpha_1^i \varepsilon(t-i).
\end{aligned}
$$

$\alpha_1^I \to 0$ as $I \to \infty$, given the process is stationary, i.e. given $|\alpha_1| < 1$. The AR(1) process may therefore be described as a MA(∞) process:

$$X(t) - \mu = \sum_{i=0}^{\infty} \alpha^i \varepsilon(t-i). \qquad (3.18)$$

By considering the first moments of a stationary AR(p) process (see Granger and Newbold (1986), p.18), it is worth noting that the variance and covariances are the solutions of $p + 1$ equations which have to be solved simultaneously:

$$E[x(t)] = \mu,$$

$$\lambda(0) = \sum_{i=1}^{p} \alpha_i \lambda(i) + \sigma^2,$$

$$\lambda(\tau) = \sum_{i=1}^{p} \alpha_i \lambda(\tau - i) + \sigma^2 \qquad \tau = 1, 2, \cdots$$

(3.19)

For the AR(1) process, the second moments are given with:

$$\lambda(0) = \frac{\sigma^2}{1 - \alpha_1^2},$$

$$\lambda(\tau) = \frac{\sigma^2 \alpha_1^\tau}{1 - \alpha_1^2}.$$

(3.20)

For positive α_1 the autocovariance function decays exponentially in τ. When α_1 is negative the autocovariance function declines also exponentially but oscillates between negative and positive values and hence adjacent observations are negatively correlated. Contrary to the MA(q) model, the autocovariance of a AR(p) process decays smoothly towards zero as the number of lags increases. Thus, the process exhibits infinite memory.

Autoregressive-Moving Average Process An ARMA (p, q) process, which includes both the AR(p) and the MA(q) processes as special cases, is given with:

$$X(t) - \mu = \sum_{i=1}^{p} \alpha_i (X(t - i) - \mu) + \sum_{j=0}^{q} \beta_j \varepsilon(t - j),$$

(3.21)

where $\beta_0 = 1$, $\alpha_p \neq 0$, $\beta_q \neq 0$.

Hence, $X(t) - \mu$ is represented as a linear function of lagged values and innovations. The acronym ARMA (p, q) stands for *autoregressive moving average process* of orders p and q.

The stationary condition for an ARMA(p, q) process is the same as that of an AR(p) model (see (3.14)). The advantage of the ARMA modelling lies in the fact that a stationary time series may often be described by an ARMA model involving fewer parameters than a pure MA or AR process. Therefore the ARMA model allows for a parsimonious representation of time series. A formula for the variance and covariances of the general ARMA(p, q) is rather complicated (see for instance Hamilton (1994), p.60). Following moments can be calculated for the ARMA(1,1) process (see Granger and Newbold (1986), p.27):

$$E[x(t)] = \mu,$$

$$\lambda(0) = \frac{(1 + 2\alpha_1\beta_1 + \beta_1^2)\sigma^2}{1 - \alpha^2},$$

$$\lambda(1) = \frac{(1 + \alpha_1\beta_1)(\alpha_1 + \beta_1)\sigma^2}{1 - \alpha^2}, \qquad (3.22)$$

$$\lambda(\tau) = \alpha_1\lambda(\tau - 1) \, , \, \tau \geq 2.$$

Thus, the covariance function begins at some starting value $\lambda(1)$, which is a function of both α_1 and β_1, and then decays geometrically from $\lambda(1)$. Hence, the ARMA(p, q) process inherits the infinite memory property of the AR(p) process.

Properties of ARMA Processes The ARMA approach provides models for the conditional mean of a time series. Let's consider the AR(1) where the innovations are supposed to be independent:

$$X(t) - \mu = \alpha_1(X(t - 1) - \mu) + \varepsilon(t). \qquad (3.23)$$

Given the realisation $x(t - 1)$ of $X(t - 1)$, the conditional mean of $X(t)$ is

$$\alpha_1(x(t - 1) - \mu) + \mu, \qquad (3.24)$$

and, depending on the value of $x(t - 1)$, may differ from the unconditional mean μ. Hence, the conditional mean can change through time and therefore allows for non-trivial prediction of future $X(t)$. The conditional variance reduces to the variance of $\varepsilon(t)$ and is therefore equal to σ^2, while the unconditional variance is $\frac{\sigma^2}{1-\alpha_1^2}$ (see equation(3.20)). More generally, if the innovations ε are independent the conditional variance of the process at time t depends only upon the last innovation $\varepsilon(t)$ and is therefore constant. Constant variance is generally termed *homoskedasticity* and hence the model is said to exhibit conditional homoskedasticity.

The Wold's decomposition theorem suggests that any weak stationary process can adequately be approximated by a linear deterministic process plus some stationary ARMA process. It is an obvious reason for the wide use of ARMA models in times series analysis. However, the Wold representation requires fitting an infinite number of parameters ψ_j to the finite number of observations. Moreover, as (3.22) shows, ARMA processes, whose lagged value parameters α_i and lagged innovation parameters β_i are non-zero, exhibit autocorrelation. Thus, only the ARMA (0,0) white noise process is not correlated. This reflects the basic property of a white noise series that earlier terms contain no information about later terms. If a series of observation indicates white noise, there is no reason to fit an ARMA model. It will however be shown below, that non-trivial models produce white noise.

Stochastic Processes for the Risk Factors

Consider an univariate time series of a risk factor whose prices are evaluated at discrete times. $w(t)$ denotes the price of the risk factor at time t. If $w(t)$ represents the price of a stock, possible dividends paid between $t - 1$ and t are assumed to be reinvested in the stock. Finance theory suggests that price series grow roughly exponentially with time. As a result, the series may have a trend and its variance is proportional to the price level. Therefore the process $\{w(t)\}$ can not generally be described as a stationary time series, and a transformation is needed to bring it back to a stationary model.

Considering the first differences $\hat{w}(t)$ is a straightforward way to remove the trend:

$$\hat{w}(t+1) = w(t+1) - w(t). \tag{3.25}$$

Differencing the $w(t)$ may however produce a series that is stationary in mean but not in variance as the variance of $\hat{w}(t)$ remains proportional to the price level. In order to stabilise the variance the series of *single returns* $\{R^*(t, t+1)\}$ can be considered:

$$R^*(t, t+1) = \frac{w(t+1)}{w(t)} - 1. \tag{3.26}$$

Single returns have the further advantage to allow comparison between different series. They are however difficult to handle for time aggregation purposes:

$$
\begin{aligned}
R^*(t, t+j) &= \frac{w(t+j)}{w(t+j-1)} \cdots \frac{w(t+1)}{w(t)} - 1 \\
&= [(1 + R^*(t+j-1, t+j)) \cdots (1 + R^*(t, t+1))] - 1 \\
&= \left[\prod_{k=1}^{j} (1 + R^*(t+k-1, t+k)) \right] - 1.
\end{aligned} \tag{3.27}
$$

To overcome the problem the *logarithmic price changes (log-returns, compound returns)* $R(t, t+1)$ may be used:

$$R(t, t+1) = \log\left(\frac{w(t+1)}{w(t)} \right). \tag{3.28}$$

The log-return over j periods reduces to the sum of the j single-period returns:

$$
\begin{aligned}
R(t, t+j) &= \log\left(\frac{w(t+j)}{w(t+j-1)} \cdots \frac{w(t+1)}{w(t)} \right) \\
&= \log\left(\frac{w(t+j)}{w(t+j-1)} \right) + \cdots + \log\left(\frac{w(t+1)}{w(t)} \right) \\
&= R(t+j-1, t+j) + \cdots + R(t, t+1).
\end{aligned} \tag{3.29}
$$

Obviously, the price $w(t+j)$ can be regained from the returns and the initial price $w(t)$ by:

$$w(t+j) = w(t)\exp\left[\sum_{i=1}^{j} R(t+i-1,t+i)\right].\qquad(3.30)$$

Note that the *law of limited liability* states that risk factor prices should never be negative and requires therefore price distributions with non-negative support. Log-returns offer the additional advantage that candidate distributions (for the log-returns) may exhibit infinite support. The transformation shown in (3.30) ensures that the law of limited liability will not be violated. The log-return over one period, $R(t-1,t)$, will be denoted by $R(t)$ in the following.

3.2 The Economic Assumption: the Efficient Market Hypothesis

Economical considerations about the functioning of the financial markets give some insight about how financial prices move. A general working hypothesis is the informational efficiency of the capital markets. In an efficient capital market, prices adjust instantaneously to the arrival of new information and therefore current prices reflect all available relevant information. Three assumptions are made which imply efficient capital markets (see Reilly (1994), p.195):

1. A large number of profit maximising participants analyse and value securities.
2. New information regarding securities comes to the market in a random fashion.
3. Investors adjust security prices rapidly to reflect the effect of new information.

The combined effect of information coming in a random fashion and rapid price adjustment gives rise to suppose that price changes are random and therefore not predictable. Since current prices contain all available information the best forecast of future prices does not require past prices.

The notion of market efficiency has been operationalised by Fama in two influential surveys (Fama (1970, 1976)), where three forms of market efficiency are distinguished:

1. The *weak form EMH* (efficient market hypothesis) assumes that prices fully reflect all security market information. This hypothesis implies that past returns should have no relationship with future rates of returns. If the weak form hypothesis holds, there is no gain for trading rules based on past market data (chart or technical analysis).

2. The *semistrong form EMH* asserts that prices fully reflect all public information. Public information includes market and also all non-market information like political and economic news or financial statements.

3. The *strong form EMH* assumes that stock prices fully reflect all information from public and private sources. This means that no investor should be able to consistently derive above-average profits.

Clearly, the three definitions of market efficiency differ in their specification of the information set. The strong form EMH implies the semistrong form EMH, which in turn implies the weak form EMH. The hypothesis of (weak) market efficiency suggests that there should be no recurring patterns in a returns series. Otherwise, rational investors would make use of it to derive profitable trading rules. In their efforts to derive benefit from the patterns they would cause them to disappear.

In statistical terms, the efficient market hypothesis rules out the presence of serial correlation in return series. Based on it, two main theories describing the behaviour of prices have been proposed in the literature: the *random walk model* and the *martingale hypothesis*.

In the *random walk model*, which was anticipated by Bachelier (1900), the return process is represented by a cumulated series of probabilistically independent shocks. Under the random walk hypothesis the returns can be written as:

$$R(t) = \mu + \varepsilon(t), \quad \varepsilon(t) \sim ID(0, \sigma^2). \tag{3.31}$$

where μ represents the expected return and $\varepsilon(t)$ is strict white noise, i.e. independent distributed with mean zero and variance σ^2.

Early contributions to the random walk model (for instance Osborne (1959)) assume $\varepsilon(t)$ being identically and normally distributed. This can be expected, as predicted by the central limit theorem, if the daily returns arise from a sum of a large numbers of intraday log-returns with identical distribution and finite variance. However, the sample characteristics of price series are frequently inconsistent with the Gaussian assumption as it will be discussed in the next section. In their attempt to reconcile theory with the empirical evidence Mandelbrot (1963b) and Fama (1965) propose non-Gaussian independent and identically distributed innovations. Granger and Morgenstern (1970), in their detailed development of the random walk, do not require the returns to be identically distributed but still maintain the independence assumption.

The *martingale hypothesis*, first introduced by Samuelson (1965) within the context of market efficiency, is less restrictive than the random walk model. A derivation of Samuelson's model with a detailed discussion of the martingale model can be found in LeRoy (1989). Samuelson shows that if prices are modelled as discounted expected cash-flows, then returns, adjusted for their expected value μ, are a martingale difference or fair game (for a formal discussion, see Sect. A.2.6):

$$E[R(t) - \mu \mid \mathcal{F}(t)] = 0, \tag{3.32}$$

where $\mathcal{F}(t)$ represents the filtration generated by the past returns at time t, i.e. the information set composed of the past returns $\{(R(t-1), R(t-2), \cdots\}$. Martingale differences are serially uncorrelated with any function $h(\mathcal{F}(t))$ of the past observations:

$$E[(R(t) - \mu)h(\mathcal{F}(t)) \mid \mathcal{F}(t)] = h(\mathcal{F}(t))E[R(t) - \mu \mid \mathcal{F}(t)] = 0, \tag{3.33}$$

which means that the martingale hypothesis rules out any dependence of the conditional expectation of $R(t) - \mu$ on the information available at time t. Thus, forecasting based on past prices are ineffective. Since martingale differences are not correlated, $\{R(t) - \mu\}$ will be white noise, given the series is stationary. Note that the random walk model, which assumes independent rather than just uncorrelated innovations is more restrictive than the martingale hypothesis. The random walk model does not only rules out dependence of the conditional expectation, but also dependence involving higher conditional moments, notably the conditional variance.

The efficient market hypothesis has been investigated for almost all important financial markets. Two main categories of tests have been formulated. The first one compares the performance of trading rules based on past market information with a buy-and-hold policy. Clearly, if the markets are efficient, no trading rule should be able to outperform the buy-and-hold policy. The second category of tests investigates the statistical consequences of market efficiency, i.e. the independence or lack of serial correlation of the returns. Since tests for independence are challenging, most studies have focussed on correlation tests and have therefore implicitly jointly tested the martingale hypothesis.

The results of the voluminous empirical research on the topic show that the individual markets do not exhibit the same level of efficiency. In large markets with high trading volumes price changes are mostly uncorrelated and therefore unpredictable. As the financial markets increase in professionalism the efficiency hypothesis, at least in its weakest form, seems to be reasonable. Many studies have shown that investors eventually can not use the small market imperfections to derive profits which overstate the transaction costs.

3.3 Empirical Evidence for the Returns

Empirical evidence indicates that distribution of the returns is not characterised by normal distribution. Following patterns have been found:

- Calendar effects
- Leptokurtosis of the daily returns and little evidence of negative skewness
- Autocorrelation of the squared returns

3.3.1 Calendar Effects

Calendar effects include the seasonal patterns that can be found in the returns of the risk factors. This has been studied in detail for stock prices. For example the January effect is well documented for the US Market. Investors tend to sell assets towards the end of the year and buy the same or other assets again after the new year for tax reasons. This could produce a negative pressure on prices at the end of the year and an upward pressure in early January. Such persistent abnormal returns are not compatible with the efficient market hypothesis. Several studies have shown that such an effect tends to vanish and that abnormal profits are generally not in excess of the transaction costs.

However, some calendar effects are still compatible with the efficient market hypothesis and actually affect the observed returns of the risk factors. Returns are recorded on calendar time scale but business time does not coincide with physical time. Most markets are closed at night and during week-ends and holiday. The returns from Friday's close to Monday's close represent for instance the result of an investment based of 72 hours rather than 24 hours. The volatility of the monday returns could therefore be higher than the volatility of returns on other days. Several studies have shown that the Monday returns have in fact a higher standard deviation than the returns of the other days. However, the ratio observed is not as high as expected if the prices follow a standard Brownian motion. In this case the Monday returns standard deviations should be about $\sqrt{3} \approx 1.73$ times the standard deviation of the returns on other days. French (1980) finds a ratio of 1.2 for the US stocks and Taylor (1986) a ratio of 1.12 for commodity prices and exchange rates.

3.3.2 Leptokurtosis and Weak Evidence of Skewness

Skewness is a measure of asymmetry of the frequency function about the mean. If the density is symmetric about its mean, the skewness is zero. Negative skewness indicates a longer tail to the right than to the left of the frequency function. The term "fat tails" means that extreme returns occur more frequently than implied by a normal distribution. The fat tails are generally accompanied by more density at the centre than predicted by the normal distribution, i.e. *leptokurtosis*. The frequency distribution exhibits therefore an higher and narrower peak and fatter tails than the normal distribution. The *kurtosis* measures the proportion of the frequency that lies in the center of the distribution. If the kurtosis is greater than three (kurtosis of the normal distribution) the centre of the distribution contains a larger proportion of the frequency than the normal.

The classical measures of skewness γ and of kurtosis κ are based on the normalised third respectively fourth moment of the distribution:

$$\gamma = E\left[\frac{(R(t) - \mu)^3}{\sigma^3}\right],$$
$$\kappa = E\left[\frac{(R(t) - \mu)^4}{\sigma^4}\right], \tag{3.34}$$

where μ is the mean and σ^2 the variance of the distribution.

Their empirical estimates from a sample of N observations r_n are given with (see Jobson (1991), pp.48,54):

$$\hat{\gamma} = \frac{N}{(N-1)(N-2)} \sum_{n=1}^{N} \frac{(r_n - \bar{r})^3}{s^3}$$
$$\approx \frac{1}{N} \sum_{n=1}^{N} \frac{(r_n - \bar{r})^3}{s^3} \text{ for large } N,$$
$$\hat{\kappa} = \frac{N(N+1)}{(N-1)(N-2)(N-3)} \sum_{n=1}^{N} \frac{(r_n - \bar{r})^4}{s^4} \tag{3.35}$$
$$\approx \frac{1}{N} \sum_{n=1}^{N} \frac{(r_n - \bar{r})^4}{s^4} \text{ for large } N,$$

where $\bar{r}$ and s^2 are the usual sample estimates of the mean respectively of the variance.

Table 3.1 displays sample statistics for a range of daily returns of risk factors as found in Duffie and Pan (1997). The returns exhibit weak evidence of negative skewness but are significantly leptokurtic. This phenomenon has been reported at many places in the literature (for stock prices see for instance Fama (1965); Praetz (1972); Blattberg and Gonedes (1974); Kon (1984); Tucker (1992); Kim and Kon (1994); Hurst and Platen (1997), for exchange rates see Boothe and Glassman (1987); Tucker and Pond (1988); Huissman et al. (1998)).

However, it should be noted that the leptokurtosis of the returns is particularly significant for the daily returns. The evidence of non-normality is much weaker for longer horizon (monthly) returns (see Blattberg and Gonedes (1974)).

Stock	s(annual)	$\hat{\gamma}$	$\hat{\kappa}$
S&P 500	15.90 %	-4.8	110.7
Nasdaq	15.20 %	-5.1	109.7
NYSE All Share	14.70 %	-5.2	121.4
Mexico Bolsa (pesos)	26.30 %	-0.2	7.8
Mexico Bolsa (U.S $)	32.00 %	0.0	14.9
FTSE 100	15.00 %	-1.7	28.6
FTSE All Share	13.60 %	-1.9	29.1
German DAX 30 Perf	19.70 %	-0.9	15.6
France DS Market	17.50 %	-0.9	13.1
France CAC 40	19.40 %	-0.6	10.9
Sweden Veckans Aff.	18.40 %	-0.4	12.1
Milan B.C.I.	20.30 %	-0.9	13.3
Swiss Performance	15.60 %	-2.6	32.0
Australia All Ordinaries	17.10 %	-7.8	198.3
Nikkei 500	20.40 %	-0.4	34.8
Hang Seng	26.70 %	-6.4	143.6
Bangkok S.E.T.	25.00 %	-0.6	9.8
Taiwan Weighted	36.10 %	-0.1	5.0
U.S.$: English Pound	11.00 %	-0.2	5.9
U.S.$: Mexican New Peso	18.90 %	-8.1	217.5
U.S.$: German Mark	11.30 %	-0.1	5.3
U.S.$: French Franc	10.90 %	0.0	5.7
U.S.$: Swedisch Krone (SK)	10.50 %	-0.6	10.9
U.S.$: Italian Lira	11.30 %	-0.6	8.6
U.S.$: Swiss Franc	12.60 %	0.0	4.9
U.S.$: Australian Dollar	9.40 %	-0.7	8.1
U.S.$: Japanese Yen	11.10 %	0.4	8.1
U.S.$: Hong Kong Dollar	0.80 %	-0.6	17.4
U.S.$: Thai Baht	2.40 %	0.7	33.8
U.S.$: Canadian Dollar	4.40 %	-0.1	7.2
Gold (First Nearby)	13.40 %	-0.7	11.4
Oil (First Nearby)	38.70 %	-0.8	21.8
Oil (Sixth Nearby)	27.50 %	-0.6	15.6

Table 3.1. Sample statistics for selected markets

3.3.3 The Autocorrelation of the Squared Returns

It is well known since Mandelbrot (1963b) and Fama (1965) that daily squared returns exhibit significant autocorrelation even though the returns are approximately uncorrelated. This means that large returns in either direction are likely to be followed by large returns in either direction. Squared returns are directly related to the unconditional variance of the returns:

$$\begin{aligned} \text{var}\,[R(t)] &= E\,[R(t) - E[R(t)]]^2 \\ &= E\,[R(t)^2] - (E[R(t)])^2\,. \end{aligned} \tag{3.36}$$

As Jorion (1995) notes, the average term $E[R(t)^2]$ dominates the term $(E[R(t)])^2$ by a typical factor of 700 to one for daily returns. Therefore, ignoring squared expected returns is unlikely to cause a perceptible bias in the variance estimate. As a result, autocorrelation in the variance of the returns is a plausible explanation for the autocorrelation of the squared returns. In the literature this phenomenon is often referred to as volatility clustering. The correlation between observations of a stationary series which are separated by a lag of τ is usually estimated from a sample of N observations $r(t)$ by the sample autocorrelation coefficient (see Campbell et al. (1997), p.45):

$$\hat{\rho}(\tau) = \frac{\hat{\lambda}(\tau)}{\hat{\lambda}(0)}, \tag{3.37}$$

where:

$$\hat{\lambda}(\tau) = \frac{1}{N} \sum_{t=1}^{N-\tau} (r(t) - \bar{r}_N)(r(t+\tau) - \bar{r}_N),$$

and:

$$\bar{r}_N = \frac{1}{N} \sum_{t=1}^{N} r(t).$$

Fuller (1976) has shown that the sample autocorrelations $\hat{\rho}(\tau)$ are asymptotically independent normal with zero mean and variance $1/N$ when the returns are independent and therefore uncorrelated. This result is often used to test hypotheses about the autocorrelation of a series at lag τ: at the 1% level of significance the zero correlation hypothesis for $\rho(\tau)$ is rejected if $|\sqrt{(N)}\hat{\rho}(\tau)| > 2.33$. The null hypothesis of non-zero autocorrelation at m lags is typically tested with the *portmanteau* or Q_m statistic of Box and Pierce (1970) which is based on the sum of the first m autocorrelations:

$$Q_m = N \sum_{\tau=1}^{m} \rho^2(\tau). \tag{3.38}$$

Since the sample correlations, under the zero-correlation hypothesis, are asymptotically independent and normally distributed $\mathcal{N}(0, 1/N)$, the empirical counterpart of Q_m:

$$\hat{Q}_m = N \sum_{\tau=1}^{m} \hat{\rho}^2(\tau), \qquad (3.39)$$

will be asymptotically distributed as χ_m^2 (chi square with m degrees of freedom). Hence, comparing the value of $\hat{Q}_m$ with the tabulated values of the χ_m^2 distribution allows to detect from departures from zero autocorrelations at the first m lags. For moderate sample size N, the Ljung and Box statistic provides a finite sample correction which yields a better fit to the χ_m^2 distribution (see Ljung and Box (1978)):

$$\hat{Q}_m = N(N+2) \sum_{\tau=1}^{m} \frac{\hat{\rho}^2(\tau)}{N - \tau}. \qquad (3.40)$$

3.4 Models for the Risk Factor Dynamics

The theory of market efficiency suggests processes for the log-returns series to be independent or almost uncorrelated. This fact has been largely confirmed by the empirical evidence and hence one may expect little success in trying to capture repeating patterns in the conditional mean as the ARMA models do. Candidate models for the returns should therefore produce white noise but also capture some of the other salient features of the returns like calendar effects, leptokurtosis and autocorrelation of the squared returns.

Let's consider the following class of processes for the log-returns:

$$\{R(t)\} = \{\mu(t)\} + \{V(t)\}\{\varepsilon(t)\}, \qquad (3.41)$$

where $\{\mu(t)\}$ is the drift process, $\{V(t)\}$ a non-negative scaling process and $\{\varepsilon(t)\}$ a standard Gaussian white noise, i.e. $\varepsilon(t)$ is strict white noise with standard normal distribution $\mathcal{N}(0,1)$. Note that $V(t)$ is an unobservable quantity which can be obtained only indirectly via the distribution of the log-returns. The models presented herein differ in the modelling of $V(t)$. The *generic model* assumes that $V(t)$ is equal to some constant σ. The *ARCH* models characterise $V(t)$ as a deterministic function of the past log-returns: $V(t) = h\left[R(t-1), R(t-2)\cdots\right]$. In the *stochastic variance* approach $V(t)$ is expressed as a function of a random variable $\eta(t)$ and of its own past: $V(t) = h\left[V(t-1), V(t-2), \cdots, \eta(t)\right]$.

All models assume $\{\varepsilon(t)\}$ and $\{V(t)\}$ being stochastically independent or at least that $\varepsilon(t)$ is independent of all $\varepsilon(t-i)$ and all $\{V(t-i)\}$, $i > 0$ (see Taylor (1986) p.75). This latter property is referred to as conditional independence.

As stated above, there is little chance of finding repeating patterns in the conditional mean and therefore most models assume that the drift process $\{\mu(t)\}$ is constant. In case $\mu(t) = \mu$, (3.41) can be rewritten:

$$\{R(t)\} - \mu = \{V(t)\}\{\varepsilon(t)\}. \tag{3.42}$$

The log-returns adjusted for the mean μ, $R(t) - \mu$, will be termed *adjusted returns* in the following.

The process $\{R(t)\} - \mu$ is uncorrelated:

$$\begin{aligned}
\operatorname{cov}\left[R(t) - \mu, R(t+\tau) - \mu\right] &= E\left[V(t)V(t+\tau)\varepsilon(t)\varepsilon(t+\tau)\right] \\
&= E\left[V(t)V(t+\tau)\varepsilon(t)\right]E\left[\varepsilon(t+\tau)\right] \\
&= 0,
\end{aligned} \tag{3.43}$$

and therefore white noise, given it is stationary.

Since $V(t)$ and $\varepsilon(t)$ are conditionally independent, the mean and the variance of the log-return conditional to a realisation $v(t)$ of $V(t)$ are given with:

$$E\left[R(t) - \mu \mid v(t)\right] = v(t)E\left[\varepsilon(t)\right] = 0. \tag{3.44}$$

$$\begin{aligned}
\operatorname{var}\left[R(t) - \mu \mid v(t)\right] &= E\left[V^2(t)\varepsilon(t)^2 \mid v(t)\right] \\
&= v(t)^2 E\left[\varepsilon(t)\right] \\
&= v(t)^2.
\end{aligned} \tag{3.45}$$

Hence, the conditional variance of the return is equal to $v(t)^2$ and therefore $\{V(t)\}$ is often referred to as the *conditional volatility* process. The ARCH and stochastic variance models allow for changing the conditional volatility. The conditional volatility is therefore not a parameter but a process which evolves over time. This property is referred to as conditional heteroskedasticity.

Economically, a change in the conditional volatility can be interpreted as the change in the level of market activity. In turn the market activity is influenced by the amount and importance of new information, the trading volume, the number of active traders, the interest in the market relative to others and also seasonal factors (see Taylor (1986), p.71).

3.4.1 The Generic Model for the Log-returns

In this model $V(t)$ is equal to some constant σ and hence

$$R(t) - \mu = \sigma\varepsilon(t). \tag{3.46}$$

Obviously, the log-returns are independent and normally distributed $\mathcal{N}$ (μ, σ^2) and therefore follow a random walk. Consequently, the risk factor prices will be log-normally distributed. The assumption of constant variance implies that information which causes price changes comes in an uniform fashion, that is the level of activity of the market is constant. The generic model can therefore not accommodate volatility clustering or leptokurtosis. It plays, however, a fundamental role in finance. The mean-variance analysis and the Capital Asset Pricing Model are based on normal distributed risk factors and assume implicitly returns with constant conditional variance. The continuous-time equivalent of the generic model, the geometric Brownian Motion, describes the dynamics of risk factors in most option pricing models and will be discussed in Sect. 3.6. One major advantage of the generic model is mathematical attractiveness:

- Mean and variance completely describe the univariate distribution.
- The multidimensional extension is straightforward:
  - As in the univariate case the multivariate normal distribution is completely defined by its first and second moments.
  - The marginal distribution of any one variable from the multivariate normal is univariate normal.
  - Every nontrivial linear combination of the components of a vector with multivariate normal distribution has a univariate normal ditribution.

3.4.2 ARCH Models

The term ARCH refers to *autoregressive conditional heteroskedasticity*. This class of model was introduced by Engle in his seminal paper (Engle (1982)). The approach assumes the conditional volatility being a deterministic function of past returns:

$$V(t) = h\left[R(t-1), R(t-2)\cdots\right], \qquad (3.47)$$

and therefore:

$$R(t) - \mu = h\left[R(t-1), R(t-2)\cdots\right]\varepsilon(t). \qquad (3.48)$$

The processes $\{V(t)\}$ and $\{\varepsilon(t)\}$ are not independent but $\varepsilon(t)$ is independent of all past values $\varepsilon(t-i)$ and $V(t-i)$, $i > 0$. As a result, the log-returns are uncorrelated and $R(t) - \mu$ is a martingale difference (for proof see Taylor (1986), pp.76,78). Given the information set $I(t)$ composed of the past returns, the realisation of $V(t)$, $v(t)$, can be calculated. Since $\varepsilon(t) \sim \mathcal{N}(0,1)$ the adjusted returns will be conditionally normal distributed:

$$R(t) - \mu \mid I(t) \sim \mathcal{N}(0, v^2(t)). \qquad (3.49)$$

The simplest ARCH model, the ARCH(1), is defined as (Engle (1982)):

$$V(t) = \left[\alpha_0 + \alpha_1 (R(t-1) - \mu)^2\right]^{\frac{1}{2}}, \tag{3.50}$$

with $\alpha_0 > 0, \alpha_1 \geq 0$ to ensure a positive conditional volatility.

The log-returns are therefore characterised by:

$$R(t) - \mu = V(t)\varepsilon(t) = \left[\alpha_0 + \alpha_1 (R(t-1) - u)^2\right]^{\frac{1}{2}} \varepsilon(t). \tag{3.51}$$

The process is covariance stationary if, and only if, $\alpha_1 < 1$; the fourth moment is finite if $3\alpha_1^2 < 1$. Then the unconditional variance and the kurtosis are given with (see Engle (1982)):

$$\mathrm{var}\,[R(t) - \mu] = \frac{\alpha_0}{1 - \alpha_1},$$
$$\kappa = 3\frac{1 - \alpha_1^2}{1 - 3\alpha_1^2}. \tag{3.52}$$

The kurtosis exceeds 3 as $3\alpha_1^2 < 1$ and so the unconditional distribution exhibits fatter tails than the normal one.

It is natural to spread the memory of the process by including p lagged variables:

$$R(t) - \mu = \left[\alpha_0 + \sum_{i=1}^{p} \alpha_i R^2(t-i)\right]^{\frac{1}{2}} \varepsilon(t), \tag{3.53}$$

with $\alpha_i > 0$. This process is referred to as the ARCH(p) model. Defining $l(t) = V^2(t)(\varepsilon^2(t) - 1)$, the ARCH($p$) model can be rewritten in terms of the adjusted squared returns:

$$\begin{aligned}
(R(t) - \mu)^2 &= V^2(t) + (R(t) - \mu)^2 - V^2(t) \\
&= \alpha_0 + \sum_{i=1}^{p} \alpha_i R^2(t-i) + V^2(t)\varepsilon^2(t) - V^2(t) \\
&= \alpha_0 + \sum_{i=1}^{p} \alpha_i R^2(t-i) + l(t).
\end{aligned} \tag{3.54}$$

Hence, the adjusted squared returns follow an AR(p) process with innovation $l(t)$. The adjusted squared returns are modelled as a linear function of p lagged values which allows for capturing their autocorrelation. If a major market movement in either direction occurred m periods ago ($m < p$) and assuming α_m is non-zero the prediction of future adjusted square returns will be large. Thus, in the ARCH(p) model large market movements tend to follow large market movements. The order of the lag p determines the length of time on which large market movements have an impact.

The ARCH approach has become very popular in the domain of finance. The modelling of the conditional variance as a deterministic function of past returns allows to construct forecasts for the variance, which is used as a key input factor for several financial applications like option pricing or asset allocation. Several hundred research papers about ARCH models have appeared since its introduction by Engle. An excellent review of the literature can be found in Bollerslev et al. (1992). The most important extension of the ARCH approach, the GARCH model, is discussed below.

GARCH A practical problem of the ARCH(p) model, for large p, is that the non-negativity constraint on the $\alpha_i's$ is often violated when the parameters are estimated. In order to obtain more flexibility, Bollerslev proposes the *generalised ARCH model*, GARCH(p, q), by adding q autoregressive terms (Bollerslev (1986)):

$$V^2(t) = \alpha_0 + \sum_{i=1}^{p} \alpha_i (R(t-i) - \mu)^2 + \sum_{j=1}^{q} \beta_j V^2(t-j), \qquad (3.55)$$

where the inequality restrictions $\alpha_0 > 0, \alpha_i \geq 0$ and $\beta_j \geq 0$, $i, j > 0$ ensure that the conditional variance is strictly positive.

The necessary and sufficient condition for the existence of the unconditional variance is (see Bollerslev (1986)):

$$\sum_{i=1}^{q} \alpha_i + \sum_{j=1}^{p} \beta_j < 1. \qquad (3.56)$$

If condition (3.56) holds, the variance is:

$$\text{var}\,[R(t) - \mu] = \frac{\alpha_0}{1 - \sum_{i=1}^{q} \alpha_i - \sum_{j=1}^{p} \beta_j}, \qquad (3.57)$$

and the process is weak stationary. There is no expression for the fourth moment of the general GARCH(p, q) model. Although the unconditional distribution in the GARCH(p, q) model has always fatter tails than the normal distribution, it is widely recognised that the model does not capture all the excess kurtosis in the data (for a discussion see Bollerslev et al. (1992)). To fully account for leptokurtosis the GARCH model may be generalised by letting $\varepsilon(t)$ have a Student-t distribution (see Bollerslev (1987)).

The dynamic properties of the GARCH(p, q) series can also be revealed by considering the adjusted square returns:

$$(R(t) - \mu)^2 = V^2(t) + (R(t) - \mu)^2 - V^2(t)$$

$$= \alpha_0 + \sum_{i=1}^{p} \alpha_i R^2(t-i) + \sum_{j=1}^{q} \beta_j V^2(t-j) + V^2(t)\varepsilon^2(t) - V^2(t)$$

$$= \alpha_0 + \sum_{i=1}^{p} \alpha_i R^2(t-i) + \sum_{j=1}^{q} \beta_j V^2(t-j) + l(t)$$

$$= \alpha_0 + \sum_{i=1}^{p} \alpha_i R^2(t-i) + \sum_{j=1}^{q} \beta_j V^2(t-j)\varepsilon(t-j) - \sum_{j=1}^{q} \beta_j l(t-j) + l(t)$$

$$= \alpha_0 + \sum_{i=1}^{m} (\alpha_i + \beta_i) R^2(t-i) - \sum_{j=1}^{q} \beta_j l(t-j) + l(t),$$

$$(3.58)$$

where $m = \max(q,p)$ and $l(t) = V^2(t)(\varepsilon^2(t) - 1)$. Clearly, the adjusted square returns have an ARMA(m,q) representation with non-gaussian innovations $l(t)$. Thus, the GARCH(p,q) is well designed to capture the autocorrelation of the adjusted square returns. Note that the extension of ARCH to GARCH is similar to the extension of MA to ARMA and the GARCH model can be seen as an infinite order ARCH which allows for a parsimonious representation of a series. This is best shown with the GARCH(1,1) model:

$$V^2(t) = \alpha_0 + \alpha_1 \varepsilon^2(t-1) + \beta_1 V^2(t-1)$$

$$= \alpha_0 + \alpha_1 \varepsilon^2(t-1) + \beta_1(\alpha_0 + \alpha_1 V^2(t-2) + \beta_1(\alpha_0 + \alpha_1 V^2(t-3) + \beta_1(\cdots$$

$$= \frac{\alpha_0}{1 - \beta_1} + \alpha_1 (\varepsilon^2(t-1) + \beta_1 \varepsilon^2(t-2) + \beta_1 \varepsilon^2(t-3) + \cdots).$$

$$(3.59)$$

The above assumes that $\beta_1 < 1$, which is always true when the process is stationary. The empirical work has shown, that the GARCH(1,1) model is sufficient to represent the most time series since the process has infinite memory (see for instance Alexander (1996), p.242).

3.4.3 Stochastic Variance Models

The basic alternative to ARCH-type modelling consists in characterising $V(t)$ as a function of its own past and of a stochastic variable $\eta(t)$ which is not dependent on past returns (see Taylor (1986), p.71):

$$V(t) = h\left[V(t-1), V(t-2), \cdots, \eta(t)\right], \qquad (3.60)$$

and therefore:

$$R(t) - \mu = h\left[V(t-1), V(t-2), \cdots, \eta(t))\right]\varepsilon(t). \qquad (3.61)$$

Thus, past returns do not cause the conditional standard deviation, and $\{\varepsilon(t)\}$ and $\{V(t)\}$ are supposed to be stochastically independent. The independence of both series allows to calculate the unconditional moments of the adjusted log-returns:

$$E[R(t) - \mu] = E[\varepsilon(t)]E[V(t)] = 0,$$
$$\text{var}\,[R(t) - \mu] = E[(R(t) - \mu)^2] = E[\varepsilon^2(t)]E[V^2(t)] = E[V^2(t)], \quad (3.62)$$
$$E[(R(t) - \mu)^4] = E[\varepsilon^4(t)]E[V^4(t)] = 3E[V^4(t)],$$

as the fourth moment of a standard normal distribution equals three.

Moreover, the odd moments of $R(t) - \mu$ are all zero since $\varepsilon(t)$ is symmetric. The process is uncorrelated (see (3.43)) and therefore a martingale difference. It will be white noise if $E[V^2(t)]$ is finite. The kurtosis is given with:

$$\kappa_R = \frac{E\left[(R(t) - \mu)^4\right]}{E\left[(R(t) - \mu)^2\right]^2} = \frac{E\left[V^4(t)\varepsilon^4(t)\right]}{E\left[V^2(t)\right]^2} = \frac{\kappa_{\varepsilon(t)}E\left[V^4(t)\right]}{E\left[V^2(t)\right]^2} = \frac{3E\left[V^4(t)\right]}{E\left[V^2(t)\right]^2}.$$
$$(3.63)$$

Since all $V(t)$ are positive $E\left[V^4(t)\right] > E\left[V^2(t)\right]^2$ if $V(t)$ has non-zero variance, i.e. is stochastic. Hence, the stochastic nature of $V(t)$ makes the distribution leptokurtic. Moreover, the squared adjusted returns can exhibit autocorrelation (see Taylor (1986), p.73):

$$E[(R(t) - \mu)^2(R(t + \tau) - \mu)^2] = E[V^2(t)V^2(t + \tau)]E[\varepsilon^2(t)\varepsilon^2(t + \tau)]$$
$$= E[V^2(t)V^2(t + \tau)], \quad (3.64)$$

and so:

$$\text{cov}[(R(t) - \mu)^2, (R(t + \tau) - \mu)^2] = E[(R(t) - \mu)^2(R(t + \tau) - \mu)^2]$$
$$- E[(R(t) - \mu)^2]^2 \quad (3.65)$$
$$= \text{cov}[V^2(t), V^2(t + \tau)].$$

Thus, the autocovariances of $(R(t) - \mu)^2$ are related to the autocovariances of $V^2(t)$. Hence, if the process of conditional variance $V^2(t)$ is positively autocorrelated, the squared adjusted log-returns will also be autocorrelated.

Discrete Mixture of Normal Distributions

The simplest conditional variance model is characterised by:

$$V(t) = \eta(t), \quad (3.66)$$

where $\eta(t)$ is a discrete random variable. As a result, the process $\{V(t)\}$ is discrete state, i.e. allows only a finite number of values. Let's consider the simplest case where $\eta(t)$ is either σ_1 or σ_2 with probability α_1 respectively α_2. Hence, the adjusted log-returns have the following form:

$$R(t) - \mu = \begin{cases} \sigma_1 \varepsilon(t) & \text{with probability } \alpha_1 \\ \sigma_2 \varepsilon(t) & \text{with probability } \alpha_2. \end{cases} \tag{3.67}$$

Since $\varepsilon(t)$ is normally distributed, the distribution of $R(t) - \mu$ becomes a *discrete mixture of normal distributions*. The calculation of its second and fourth moments for $\alpha_1 = \alpha_2 = 1/2$ is straightforward:

$$\text{var}\left[R(t) - \mu\right] = E\left[\varepsilon(t)^2\right] E\left[V(t)^2\right] = \frac{1}{2}\left(\sigma_1^2 + \sigma_2^2\right),$$

$$E\left[R(t)^4\right] = E\left[\varepsilon(t)^4\right] E\left[V(t)^4\right] = \frac{3}{2}\left(\sigma_1^4 + \sigma_2^4\right), \tag{3.68}$$

and $R(t) - \mu$ has following kurtosis:

$$\begin{aligned} \kappa &= 3\frac{2(\sigma_1^4 + \sigma_2^4)}{(\sigma_1^2 + \sigma_2^2)^2} \\ &= 3\left\{1 + \frac{(\sigma_1^2 - \sigma_2^2)^2}{(\sigma_1^2 + \sigma_2^2)^2}\right\} > 3. \end{aligned} \tag{3.69}$$

The idea of discrete mixture of normal distributions has been used in the literature in order to describe discontinuous shifts in the parameters of the return distributions. Such shifts arise from anomalies of the market like calendar effects or from the arrival of unexpected and important information. The compound normal distribution model mixes a finite number k of distributions and the mix-jump process, the superimposition of a Gaussian random walk and an independent compound Poisson process, can be described as an infinite mixture of normal distributions. Note that calendar effects may cause a shift not only in the conditional variance but also in the conditional expectation. As a result, both models allow $\mu(t)$ to vary.

Compound Normal Distribution The *compound normal distribution* model is an extension of the discrete mixture which allows shifts in the trend process $\{\mu(t)\}$. The n log-returns considered are supposed to be generated by one of the following equations (see Kon (1984)):

$$\begin{aligned} R(t) &= \mu_1 + \sigma_1^2 \varepsilon(t) \ t \in I_1, \\ R(t) &= \mu_2 + \sigma_2^2 \varepsilon(t) \ t \in I_2, \\ &\cdots \\ R(t) &= \mu_k + \sigma_k^2 \varepsilon(t) \ t \in I_k, \end{aligned} \tag{3.70}$$

where I_i $(i = 1, 2 \cdots, k)$ are homogeneous information sets with n_i observations in each set and $\sum_{i=1}^{k} n_i = n$.

All log-returns generated by an information set I_i are normally distributed with given parameters μ_i and σ_i^2. With probability α_i ($\sum_{i=1}^{k} \alpha_i = 1$) a single return is generated by the information set I_i. As a result, the unconditional distribution of $R(t)$ is a mixture of k normal distributions with different means and variances.

The model is tractable if the number of shifts is relatively small with respect to the number of observations ($k \ll n$). It is therefore likely to describe samples well which exhibit repeating patterns. Typical examples of repeating sequences are the calendar effects. For example I_1 could be the set of information occurring during the week-end and on Monday and generating the Monday returns, with higher variance than returns on other days of the week. I_2 would therefore be the information set generating the returns on the other days of the week. The model can be extended to allow the modelling of the January effects or firm specific information distributions like the disclosure of firm's earnings.

Kon (1984) fits this model for each of the 30 stocks composing the Dow Jones Industrial Average and also for three US stock market indexes. He considers the five potential model specifications $k = 1, 2, 3, 4$ or 5 and estimates the parameters with a maximum likelihood algorithm. He finds that the returns are best described as a mixture of four normal distributions for 11 stocks, a mixture of three normal distributions for 11 stocks and a mixture of two normal distributions for the remaining 12 stocks. For all indexes $k = 3$ has the better explanatory power.

Since the estimates of the α_i depend on the number of observations, the unconditional mean and variance are not constant, and therefore the model is not stationary. Moreover, for a given k the number of parameters which has to be estimated is $3k - 1$ (k means, k variances and $k - 1$ probabilities).

Mixed-Jump Processes The *mixed-jump process* consists of the superposition of a Gaussian random walk and an independent compound Poisson process with normally distributed jump amplitudes (see Press (1968)):

$$R(t) = \mu_1 + \sigma_1 \varepsilon(t) + \sum_{i=1}^{N(t)} Y_i, \qquad (3.71)$$

where $Y_1, \cdots, Y_k$ is a sequence of mutually independent random variables with normal distribution $\mathcal{N}(\mu_2, \sigma_2^2)$

$N(t)$ is a Poisson counting process with parameter λ, which represents the number of random events occurring between $t - 1$ and t. The Poisson process is widely used to count the number of times particular events occur in a given period of time. This type of modelling captures the phenomenon of a discontinuous jump in the stock price, as it might be caused by a sudden and unexpected arrival of important information, the *rare events*. Such rare events are for example the fusion of two companies or the disclosure of unexpected losses.

The continuous-time equivalent of the mixed jump process, the superposition of a Brownian motion and a Poisson process, is often used in the theory of option pricing as an alternative description of the stock price movement (see Neftci (1996); Cox and Rubinstein (1985)). Unlike the Brownian motion the size of Poisson outcomes does not depend on the time interval considered. Instead, the probability of occurrence of the outcomes is function of the time interval. In the continuous-time modelling a Poisson process allows for large (rare) events also when the time interval goes to zero.

The pdf of a random variable with Poisson distribution is:

$$f(x) = \frac{\lambda^x e^{-\lambda}}{x!}, \tag{3.72}$$

and the pdf of the returns is given with (see Kim and Kon (1994)):

$$f(r) = \sum_{n=1}^{\infty} \frac{e^{\lambda}}{n!} \lambda^n \mathcal{N}(\mu_1 + n\mu_2, \sigma_1^2 + n\sigma_2^2), \tag{3.73}$$

where $\mathcal{N}(a, b)$ denotes a normal density with mean a and variance b.

The density function is leptokurtic if $\lambda > 0$ and symmetric around μ_1 if $\mu_2 = 0$. As (3.73) shows, the model is an infinite mixture of normal densities which are mixed by Poisson probabilities. The estimation of the parameters is a challenging task which is performed through an approximation with a finite mixture of normal distributions.

Continuous Mixture of Normal Distributions

Once again the conditional variance, respectively the adjusted log-returns, are characterised by:

$$V(t) = \eta(t), \tag{3.74}$$

and:

$$R(t) - \mu = V(t)\varepsilon(t). \tag{3.75}$$

$\eta(t)$ is assumed to be a continuous random variable. The model permits therefore a continuum of values for the conditional volatility. Since $\varepsilon(t)$ is normally distributed, the distribution of $R(t) - \mu$ is a *continuous mixture of normal distribution.* The unconditional distribution of $R(t) - \mu$ is also termed *mixture distribution* in the literature while the distribution of $V(t)$ is called the *mixing distribution* (see Feller (1966), § II.5). A large review of alternative candidates for the mixing distribution is given in Hurst and Platen (1997). $V(t)$ is always assumed to be independent and identically distributed. Therefore, the model cannot account for volatility clustering but has proven to be successful in capturing the excess kurtosis of log-return series (see Praetz (1972); Blattberg and Gonedes (1974); Boothe and Glassman (1987); Hurst and Platen (1997); Huissman et al. (1998)).

Recall that $\{V(t)\}$ is a positive process since the volatility cannot be negative. The mixing distribution must therefore have non-negative support and will be positively skewed. A natural candidate for the distribution of $V(t)$ is obviously the lognormal distribution, which has been first proposed by Clark (1973). As Blattberg and Gonedes (1974) remark this model has no closed form expression for the mixture distribution which must be expressed in integral form. This presents obvious difficulties for empirical work.

Two other candidates for the modelling of mixing distribution have been widely discussed in the literature:

- the inverse of a strictly positive stable distribution,
- the inverse Gamma distribution.

In both cases the mixture distribution is known and represents a generalisation of the Gaussian model. If the mixing distribution is inverse strictly positive stable, the resultant mixture distribution is (symmetric) *stable*. In case the mixing distribution is inverse Gamma, the mixture distribution will be the *Student-t* (see Blattberg and Gonedes (1974); Praetz (1972)). The properties of both mixture distributions are discussed in the following section.

The Stable Distribution The stable distribution has been proposed by Mandelbrot (1963a,b, 1967) and Fama (1965). The symmetric stable distribution is defined generally by its log-characteristic function $\log \phi()$ since its density function is generally unknown (see Embrechts et al. (1997), p.71):

$$\log \phi(s) = \log E\left[\exp(ixs)\right] = i\delta s - |cs|^\alpha, \tag{3.76}$$

where s is a real number and $i = \sqrt{-1}$. The three parameters of the distribution are α, the characteristic exponent, c, the scale parameter, and δ, the location parameter.

A standardised symmetric-stable random variable exhibits following properties (see Tucker and Pond (1988); Blattberg and Gonedes (1974)):

1. if $\alpha < 2$
   - the tails are fatter than those of a standardised normal random variable,
   - the density function is higher in the neighbourhood of the location parameter δ than the density function of a standardised normal random variable,

2. if $\alpha = 1$ the distribution is the Cauchy distribution,

3. if $\alpha = 2$ the distribution is the standard normal distribution,

In general, all moments of order $r < \alpha$ are finite except when $\alpha = 2$, in which case all moments of all orders are finite.

The Student-t Distribution The Student-t density function with location parameter m and scale parameter $H > 0$ and degrees of freedom parameter $\nu > 0$ is (Blattberg and Gonedes (1974)):

$$f(x) = \frac{\nu^{\frac{1}{2}\nu}\Gamma\left[\frac{1}{2}(1+\nu)\right]}{\Gamma\left[\frac{1}{2}\right]\Gamma\left[\frac{1}{2}(\nu)\right]}\left[v + H(x-m)^2\right]^{-\frac{1}{2}(\nu+1)}\sqrt{H}, \qquad (3.77)$$

where $\Gamma[]$ is the gamma function.

The distribution has following properties:

1. $E[X] = m$ for $\nu > 1$,
2. $\text{var}[X] = \frac{\nu}{H(\nu-2)}$ for $\nu > 2$,
3. in general all moments of order $r < \nu$ are finite,
4. when $\nu = 1$ the distribution is the Cauchy distribution,
5. when $\nu \to \infty$ the Student distribution converges to the normal distribution,
6. the density function of the standardised Student random variable exhibits fatter tails than the density function of standardized normal variable.

Comparison Between the Stable and the Student-t Distributions
Both distributions can account for the fat tails exhibited by the empirical distributions of the daily log-returns. The parameters of importance with respect to the kurtosis is α for the stable distribution and ν for the Student distribution. Following important differences between both distributions should be noted:

1. The stable distribution exhibits fatter tails than the normal distribution for $\alpha < 2$. In this case the distribution has infinite variance. If the returns are generated from an infinite variance distribution, the sample variance, which is always finite, is no more a consistent estimator and all methods based on second moments are misleading (for instance the mean-variance analysis). The Student-t distribution has tails, which are always fatter than the normal distribution. For $\nu > 2$ the Student distribution has a finite variance.

2. Of both models only the Student-t distribution hypothesis can account for the empirical evidence that monthly log-returns are almost normally distributed. From (3.29) it is obvious that monthly log-returns represent a sum of daily log-returns. It is also well known (see Embrechts et al. (1997), p.71) that the class of the stable distributions coincides with the class of all possible limit laws for sums of identically independent distributed random variables. As a result, the distribution of monthly log-returns will converge (in distribution) to a stable distribution, given they are IID. Moreover, any sum of IID variables with finite variance will converge to a stable law with characteristic exponent $\alpha = 2$, i.e. to the normal distribution (see Embrechts et al. (1997), p.75). As a result, the sum of IID variables with Student-t distribution and $\nu > 2$ will converge to the normal distribution. On the contrary, a sum of IID fat-tailed stable variables ($\alpha < 2$) will not converge to the normal distribution but will have stable distribution with the same characteristic exponent $\alpha < 2$ (see Embrechts et al. (1997), p.74).

3. The stable densities are only in a few cases expressible explicitly via elementary functions: Cauchy distribution for $\alpha = 1$, normal distribution for $\alpha = 2$ and inverse Gaussian for $\alpha = \frac{1}{2}$. For the other cases, one can use asymptotic expansions (method of Edgeworth), that is a normal density function plus an infinite weighted sum of Hermite polynomials. On the other hand, the Student-t model allows the use of well defined density functions for every integer ν.

Stochastic Volatility Models

The *stochastic volatility* approach is an extension of the continuous mixture of normals models, where the conditional volatility is not simply equal to the continuous variable $\eta(t)$, but also depends on its own past. In this approach $\eta(t)$ is assumed to be normally distributed. To ensure the positivity of the conditional volatility, the logarithm of $V^2(t)$, $s(t)$, is modelled as a function of $\eta(t)$. The most widely used stochastic volatility model is given with (see Harvey et al. (1994)):

$$\log\left(V^2(t)\right) = s(t) = \alpha_0 + \alpha_1 s(t-1) + \eta(t), \qquad (3.78)$$

where $\eta(t)$ is $\mathcal{N}(0, \sigma^2)$.

The log-returns are therefore characterised by:

$$R(t) - \mu = V(t)\varepsilon(t)$$
$$= \exp\left(\frac{s(t)}{2}\right)\varepsilon(t). \tag{3.79}$$

Equation (3.78) shows that $s(t)$ follows an AR(1) process with Gaussian innovations and will therefore be stationary if $|\alpha_1| < 1$ with (see (3.20)):

$$\sigma_s^2 = \frac{\sigma^2}{1 - \alpha_1^2}. \tag{3.80}$$

As $\varepsilon(t)$ is always stationary, $R(t) - \mu$ will be stationary if, and only if, $s(t)$ is stationary (see Harvey et al. (1994)).

Note that $\exp(s(t))$ is lognormal and the moments of $R(t) - \mu$ can be derived from the lognormal distribution. In particular:

$$\text{var}\left[R(t) - \mu\right] = \mu_s + \frac{1}{2}\sigma_s^2$$
$$\kappa = 3\exp\left(\sigma_s^2\right) \geq 3. \tag{3.81}$$

The properties of the model are best analysed by considering the logarithm of the squared adjusted returns:

$$\log\left[(R(t) - \mu)^2\right] = s(t) + \log\left(\varepsilon(t)^2\right)$$
$$= \alpha_0 + \alpha_1 s(t-1) + \eta(t) + \log\left(\varepsilon(t)^2\right). \tag{3.82}$$

Thus, the process $\left\{\log\left[(R(t) - \mu)^2\right]\right\}$ adds the white noise process $\log\left(\varepsilon(t)^2\right)$ to the AR(1) component $\{s(t)\}$. As a result, its autocorrelation function is equivalent to that of an ARMA(1,1) (see Harvey et al. (1994)). The stochastic volatility model exhibits therefore a behaviour similar to the GARCH(1,1) model and is able to capture the autocorrelation of the squared returns. Like the GARCH models, the stochastic volatility models may be generalised by letting $\varepsilon(t)$ have a Student-t distribution in order to capture the leptokurtosis of the log-returns. Hence, the stochastic volatility models share a lot of properties with the GARCH models. The major difference between both approaches is that GARCH models assume $V(t)$ to be a deterministic function of past returns while the stochastic volatility models deal with the conditional volatility as an unobserved variable, the logarithm of which is modelled as a stochastic process. It is generally recognised that stochastic volatility models are more difficult to estimate than GARCH models but offer the advantage of a natural generalisation to multivariate series (for a discussion see for instance Harvey et al. (1994) or Ruiz (1994)).

3.5 Empirical Analysis of the Returns on Swiss Stocks

The purpose of this section is to investigate the empirical properties of the daily log-returns on Swiss stocks and to compare one alternative model, the Student-t model, with the generic normal model. Firstly, statistics for the daily log-returns are provided in order to verify the empirical properties (skewness, kurtosis and volatility clustering) of the series considered. Later, the two competing models are implemented for the one- and M-dimensional cases and their descriptive ability in the data are compared by the means of maximum likelihood estimation. The last section discusses the impact of the non-normality of adjusted returns on Value-at-Risk.

3.5.1 The Data

The sample consists of the daily log-returns on 18 stocks in the SMI (Swiss Market Index). For 15 stocks there are 1197 log-returns observed from January 3, 1994 to June 30, 1998. The sample size is 873 for the stock UBSN (Union Bank of Switzerland) (records from January 3, 1995 to June 30, 1998), 750 for the stock CLN (Clariant) (records from June 30, 1995 to June 30, 1998) and 323 for CIBN (Ciba Speciality) (records from March 13, 1997 to June 30, 1998). The sample size for CIBN is rather small for inference purposes and two other stocks of the SMI (Swissair and Swiss Life), with even smaller available sample sizes, have not been considered. The raw price data (average closing price) have been obtained from Bloomberg.

3.5.2 Descriptive Statistics and Correlation

The descriptive statistics mean, standard deviation, skewness and kurtosis are displayed in table 3.2. Under the normal distribution hypothesis, the upper and lower one percentage points of the skewness distributions are -0.17 and 0.17 for the returns with sample size 1126, - 0.19 and 0.19 for the returns on UBSN, -0.21 and 0.21 for the returns on CLN and -0.32 and 0.32 for the returns on CIBN (see for instance Jobson (1991), p.48). The hypothesis is therefore rejected for all stocks. The evidence of negative skewness found in Duffie and Pan (1997) can not be confirmed for the Swiss Market Index: more than half of the log-returns (eleven of eighteen) exhibit positive skewness.

The leptokurtosis is significantly more prominent than the skewness in the sample considered: the kurtosis statistic ranges in value from 4.21 to 40.20. Under the normal distribution hypothesis the upper percentage point of the kurtosis distribution is 3.34 for the returns with sample size 1126, 3.39 for the returns on UBSN, 3,42 for the returns on CLN and 3.63 for the returns on CIBN (see for instance Jobson (1991), p.54). Thus, the kurtosis statistics

Stock	$\bar{r} * 250$	s^2	$\hat{\gamma}$	$\hat{\kappa}$
ABB	15.47%	1.28%	-0.99981409	10.6705909
ALUN	24.51%	1.24%	-0.17079686	4.94508833
BALN	21.70%	1.57%	0.17837242	5.14473298
CIBN	39.77%	2.04%	0.21218994	4.21193761
CLN	54.63%	1.76%	1.32006558	10.7148408
CSGN	18.73%	1.61%	0.3469228	7.17820945
EMS	15.75%	1.02%	-0.14914711	4.78904204
HOL	15.95%	1.51%	-0.10417853	6.64941113
NESN	20.47%	1.10%	0.16064101	4.57606645
NOV	24.35%	1.41%	1.51194091	24.1887135
NOVN	25.37%	1.37%	2.40169342	40.206368
ROG	19.03%	1.23%	-0.16041946	5.24321726
RUKN	36.42%	1.37%	0.46726062	5.61179508
SGSI	6.02%	1.69%	-0.1883225	5.43171463
SU/N	7.44%	1.66%	0.65508219	11.3437177
SMH	2.34%	1.77%	-0.25174427	6.37754103
UBSN	23.64%	1.34%	0.49271878	7.90320893
ZURN	25.60%	1.38%	0.21807295	5.07156073

Table 3.2. Sample statistics for the daily log-returns

indicate a severe deviation from normality for all stocks.

Table 3.3 gives insight in the correlation of the log-returns and squared log-returns. Using the statistic of Fuller (see Sect. 3.3.3) for the autocorrelation at lag 1, the upper and lower percentage points of the autocorrelation distribution are -0.069 and 0.069 for the returns with sample size 1126, -0.079 and 0.079 for the returns on UBSN, -0.085 and 0.085 for the returns on CLN and -0.13 and 0.13 for the returns on CIBN. Nine returns exhibit small but significant autocorrelation at the first lag and 15 squared returns are significant autocorrelated at the first lag.

The somewhat surprising result that half of the returns have significant autocorrelation at the first lag is however tempered by the Q statistic, which considers the cumulative effect of up to 20th order autocorrelation. The upper percentage point for the Q statistic is 37.57 for all stocks. The zero autocorrelation hypothesis can not be rejected for 15 out of the 18 log-returns. The evidence of autocorrelation is much stronger for the squared log-returns where the zero autocorrelation hypothesis can not be rejected for only 5 out of the 18 stocks (see table 3.3).

Stock	Returns		Squared returns	
	$\hat{\rho}(1)$	$Q(20)$	$\hat{\rho}(1)$	$Q(20)$
ABB	0.150	53.975	0.160	147.687
ALUN	0.014	19.045	0.126	80.160
BALN	0.106	39.303	0.064	91.334
CIBN	-0.042	25.486	0.132	25.212
CLN	0.116	39.895	0.356	100.178
CSGN	0.023	20.859	0.153	70.080
EMS	0.054	32.085	0.136	52.755
HOL	0.092	31.634	0.241	176.055
NESN	0.110	44.747	0.157	132.639
NOV	0.017	18.637	0.013	6.067
NOVN	0.040	18.000	0.024	4.086
ROG	0.095	37.417	0.091	115.308
RUKN	0.131	51.161	0.149	101.786
SGSI	-0.064	24.907	0.203	76.711
SU/N	0.037	20.841	0.236	93.020
SMH	0.095	33.770	0.031	10.540
UBSN	0.093	25.550	0.079	32.684
ZURN	0.006	23.946	0.079	80.930

Table 3.3. Autocorrelations statistics for the daily log-returns

3.5.3 Implementation of an Alternative Model

The results for the Swiss stocks confirm widely the results documented in the empirical literature and justify the use of an alternative model for the dynamics of the risk factors. The unconditional Student-t model has been chosen for the following reasons:

1. The continuous mixture of normal distribution models account explicitly for leptokurtosis.
2. Under these models, the Student-t model has the best theoretical properties (convergence to normality and well defined distribution function) and has shown good empirical results (see Praetz (1972); Blattberg and Gonedes (1974); Boothe and Glassman (1987); Hurst and Platen (1997); Huissman et al. (1998)).
3. The extension in the multivariate case is straightforward.
4. The model is parsimonious, i.e. requires few parameters.

However, the assumption of independent identically Student-t distributed risk factors does not permit to capture volatility clustering. An improvement could be the use of an extension of the GARCH or stochastic volatility model with Student-t innovations.

The Estimation of the Parameters

The estimation of the parameters for the normal and Student distribution is done with the method of maximum likelihood which is widely used for fitting probability laws to data.

Consider a sample of returns $r_1, \cdots, r_n$ as a realisation of the random variables $R(1), \cdots, R(n)$ whose joint distribution $f(r_1, \cdots, r_n | \theta)$ depends on the unknown parameter or vector of parameters θ. Since all $R(i)$ are supposed to be independent and identically distributed their joint distribution is $f(r_1 | \theta) \cdots f(r_1 | \theta)$ (see Rice (1995), p.253 for instance).

The likelihood of θ as a function of $r_1, \cdots, r_n$ is defined as:

$$L(\theta) = f(r_1, \cdots, r_n | \theta). \tag{3.83}$$

The method of maximum likelihood considers the joint density as a function of θ and the maximum likelihood estimate of θ is the vector of values that maximises the likelihood function. In practice, in order to avoid the overflow of computer representations, it is more convenient to work with the *log-likelihood* $l(\theta)$:

$$
\begin{aligned}
l(\theta) &= \log \prod_{i=1}^{n} f(r_i | \theta) \\
&= \sum_{i=1}^{n} \log \left[f(r_i | \theta) \right].
\end{aligned}
\tag{3.84}
$$

Since the logarithm is a monotonic function, the maximisation of the log-likelihood is equivalent to the maximisation of the likelihood itself.

The One-Dimensional Case For the normal distribution the density function depends on the two parameters μ and σ with $\sigma > 0$:

$$f(r) = \frac{1}{\sigma \sqrt{2\pi}} \exp \left(\frac{-(r - \mu)^2}{2\sigma^2} \right). \tag{3.85}$$

The likelihood function is therefore:

$$L(\mu, \sigma) = \prod_{i=1}^{n} \frac{1}{\sigma \sqrt{2\pi}} \exp \left(-\frac{1}{2} \left[\frac{r_i - \mu}{\sigma} \right]^2 \right), \tag{3.86}$$

and the log-likelihood function is given with:

$$l(\mu, \sigma) = -n \log \sigma - \frac{n}{2} \log 2\pi - \frac{1}{2\sigma^2} \sum_{i=1}^{n} (r_i - \mu)^2. \tag{3.87}$$

The partial derivatives with respect to the parameters are:

$$\frac{\partial l}{\partial \mu} = \frac{1}{\sigma^2} \sum_{i=1}^{n} (r_i - \mu),$$

$$\frac{\partial l}{\partial \sigma} = -\frac{n}{\sigma} + \sigma^{-3} \sum_{i=1}^{n} (r_i - \mu)^2. \tag{3.88}$$

The maximum likelihood estimate of the mean $\bar{r}$ is obtained by setting the first partial equal to zero and solving the corresponding equation:

$$\hat{\mu} = \bar{r}. \tag{3.89}$$

Correspondingly, the maximum likelihood estimate $\hat{\sigma}$ is calculated by setting the second partial equal to zero and substituting μ with $\bar{r}$:

$$\hat{\sigma} = \sqrt{\frac{1}{n} \sum_{i=1}^{n} (r_i - \bar{r})^2}. \tag{3.90}$$

Thus, the maximum likelihood estimates of the mean and standard deviation for the normal distribution are equal to the sample mean and sample standard deviation.

For the Student-t model, the density function is given by:

$$f(r|m, H, \nu) = \frac{\Gamma\left(\frac{1}{2} + \frac{1}{2}\nu\right)}{\Gamma\left(\frac{1}{2}\right)\Gamma\left(\frac{1}{2}\nu\right)} \frac{\nu^{\frac{1}{2}\nu}\sqrt{H}}{[\nu + H(x - m)^2]^{\frac{1}{2}(\nu+1)}}. \tag{3.91}$$

The likelihood function is therefore:

$$L(m, H, \nu) = \prod_{i=1}^{n} \frac{\Gamma\left(\frac{1}{2} + \frac{1}{2}\nu\right)}{\Gamma\left(\frac{1}{2}\right)\Gamma\left(\frac{1}{2}\nu\right)} \frac{\nu^{\frac{1}{2}\nu}\sqrt{H}}{[\nu + H(r_i - m)^2]^{\frac{1}{2}(\nu+1)}}, \tag{3.92}$$

and the log-likelihood function is given with:

$$l(m, H, \nu) = \sum_{i=1}^{n} \log \left[\frac{\Gamma\left(\frac{1}{2} + \frac{1}{2}\nu\right)}{\Gamma\left(\frac{1}{2}\right)\Gamma\left(\frac{1}{2}\nu\right)} \frac{\nu^{\frac{1}{2}\nu}\sqrt{H}}{[\nu + H(r_i - m)^2]^{\frac{1}{2}(\nu+1)}} \right]$$

$$= \sum_{i=1}^{n} \left[\log \Gamma \left[\frac{1}{2} + \frac{1}{2}\nu\right] - \log \left[\Gamma\left(\frac{1}{2}\right)\Gamma\left(\frac{1}{2}\nu\right)\right] + \frac{1}{2}\nu \log \nu \right. \tag{3.93}$$

$$\left. + \frac{1}{2}\log(H) - \frac{1}{2}(\nu + 1)\log\left[\nu + H(r_i - m)^2\right] \right].$$

The search of the maximum likelihood estimators for the Student-t distribution is a multidimensional non-linear optimisation problem. This is done by using numerical methods. A quasi-Newton method, the Davidon-Fletcher-Powell algorithm, has been implemented as described in Press et al. (1992), pp.426-429. The method proceeds by iteration and requires starting values for the parameters. Moreover, it is well known that the success of iterative optimisation methods depends crucially on the starting parameters. Since a good guess for the parameter values is a priori not available, an intensive grid search has been performed, i.e. the log-likelihood function has been valuated for 8'000'000 sets of parameters. From these 8'000'000 sets, the 1000 sets with the higher likelihood value have been used as starting parameters for an optimisation based on the Davidon-Fletcher-Powell algorithm. This rather computer intensive procedure is necessary since a single optimisation leads to a local but not necessarily to the global maximum. The Davidon-Fletcher-Powell algorithm requires the computation of the first partial derivatives of the log-likelihood function at arbitrary points. These partial derivatives are:

$$
\begin{aligned}
\frac{\partial l}{\partial m} &= \sum_{i=1}^{n} -\frac{1}{2}(\nu+1)\frac{-2H(r_i-m)}{\nu+H(r_i-m)^2} \\
&= H(\nu+1)\sum_{i=1}^{n}\frac{(r_i-m)}{\nu+H(r_i-m)^2}, \\
\frac{\partial l}{\partial H} &= \sum_{i=1}^{n}\left[\frac{1}{2H}-\frac{\nu+1}{2}\frac{(r_i-m)^2}{\nu+H(r_i-m)^2}\right] \\
&= -\frac{n(\nu+1)}{4H}\sum_{i=1}^{n}\frac{(r_i-m)^2}{\nu+H(r_i-m)^2}, \\
\frac{\partial l}{\partial \nu} &= \sum_{i=1}^{n}\left[\frac{1}{2}\left(\psi(\frac{1}{2}+\frac{\nu}{2})-\psi(\frac{\nu}{2})+\log(\nu)\right.\right. \\
&\quad \left.\left. -\log(\nu+H(r_i-m)^2)-\frac{\nu+1}{\nu+H(r_i-m)^2}\right)\right] \\
&= \frac{n}{2}\left[\psi(\frac{1}{2}+\frac{\nu}{2})-\psi(\frac{\nu}{2})+\log(\nu)\right]
\end{aligned}
\tag{3.94}
$$

where $\psi(z) = \frac{\partial \log \Gamma(z)}{\partial z} = \frac{\Gamma'(z)}{\Gamma(z)}$.

Three alternatives have been calculated:

- *Alternative 1*
  All three parameters can move freely inside the allowed range ($m \in \mathbb{R}, H \in \mathbb{R}_+, \nu \in \mathbb{R}_+$) for the search of the optimum.
- *Alternative 2*
  The location parameter is kept fixed with estimated value equal to the sample estimate: $\hat{m} = \bar{r}$. ν can vary and $\hat{H}$ is a function of ν and the sample variance: $\hat{H} = \frac{\nu}{(\nu-2)\hat{\sigma}^2}$.

- *Alternative 3*

  The estimates of the location and scale parameters are the same as in alternative 2. The estimates of the number of degrees of freedom may only take integer values.

Table 3.4 displays the maximum log-likelihood estimators for the normal distribution model and for the three alternative Student-t distribution models with the corresponding estimates of the number of degrees of freedom. The estimated numbers of degrees of freedom range from 2.08 to 8.68 (15 out of 18 are between 3.75 and 5.75) and therefore indicate that all returns have finite variance ($\nu > 2$) but are highly non-normally distributed.

Stock	Student (1) log-like.	df	Student (2) log-like.	df	Student (3) log-like.	df	Normal log-like.
ABB	3'383.69	4.25	3'382.92	4.35	3'382.40	4.00	3'310.69
ALUN	3'365.29	8.68	3'365.27	8.48	3'365.23	9.00	3'346.82
BALN	3'116.92	5.44	3'116.63	5.61	3'116.51	6.00	3'082.95
CIBN	806.17	5.54	806.10	5.87	806.09	6.00	799.42
CLN	2'384.67	2.08	2'374.24	3.00	2'374.24	3.00	2'170.86
CSGN	3'124.69	3.89	3'124.44	4.15	3'124.34	4.00	3'054.68
EMS	3'606.17	4.55	3'605.92	4.91	3'605.90	5.00	3'570.26
HOL	3'191.36	4.08	3'191.18	4.26	3'190.88	4.00	3'126.43
NESN	3'504.23	6.20	3'504.05	6.32	3'503.98	6.00	3'477.98
NOV	3'307.50	4.41	3'307.21	4.12	3'307.13	4.00	3'201.18
NOVN	3'394.60	3.76	3'394.27	3.46	3'391.81	4.00	3'232.84
ROG	3'389.65	5.75	3'389.46	5.82	3'389.43	6.00	3'355.23
RUKN	3'271.77	5.05	3'271.36	5.25	3'271.28	5.00	3'231.97
SGSI	3'041.72	4.29	3'041.24	4.62	3'040.97	5.00	2'994.73
SMH	2'999.04	4.13	2'998.83	4.42	2'998.15	4.00	2'944.25
SU/N	3'087.05	4.37	3'086.98	4.37	3'086.44	4.00	3'014.39
UBSN	2'563.59	5.38	2'563.56	5.34	2'563.44	5.00	2'526.73
ZURN	3'264.83	4.88	3'264.70	5.12	3'264.68	5.00	3'226.55

Table 3.4. Maximum likelihood estimators for the daily log-returns

All three Student-t models outperform the normal model for all stocks. This is not surprising as the normal distribution is a special case of the Student-t distribution. Note that both models are nested and therefore a test to discriminate between both hypotheses may be constructed. The *likelihood ratio test* Λ (see Rao (1973)) is given with:

$$\Lambda = \frac{L_{\text{normal}}}{L_{\text{Student-t}}}. \tag{3.95}$$

The asymptotic distribution of $-2\log\Lambda$ is chisquare with degrees of freedom equal to the difference in the number of parameters between the two models. The larger the value of $-2\log\Lambda$ the better the Student-t model is. As the number of parameters to estimate is three for the Student-t distribution and two for the normal model, the distribution of the likelihood ratio test is chisquare with one degree of freedom $\left(\chi^2_{\{1\}}\right)$.

The upper one percentage point of the $-2\log\Lambda$ is therefore 6.63. It is interesting to note that the value of this statistic increase with the number of degrees of freedom of the chisquare distribution and therefore penalises models which require many parameters to estimate. Table 3.5 displays the statistics $-2\log\Lambda$ for the three alternative models. The three models are for all stocks significantly better than the normal model.

Stock	$-2\log\Lambda$ alternative 1	$-2\log\Lambda$ alternative 2	$-2\log\Lambda$ alternative 3
ABB	146.00	144.46	143.42
ALUN	36.94	36.90	36.82
BALN	67.94	67.36	67.12
CIBN	13.50	13.36	13.34
CLN	427.62	406.76	406.76
CSGN	140.02	139.52	139.32
EMS	71.83	71.33	71.29
HOL	129.86	129.50	128.90
NESN	52.50	52.14	52.00
NOV	212.64	212.06	211.90
NOVN	323.52	322.86	317.94
ROG	68.84	68.46	68.40
RUKN	79.60	78.78	78.62
SGSI	93.98	93.02	92.48
SMH	109.58	109.16	107.80
SU/N	145.32	145.18	144.10
UBSN	73.72	73.66	73.42
ZURN	76.56	76.30	76.26

Table 3.5. $-2\log\frac{L_{\text{normal}}}{L_{\text{alternatives}}}$ statistics

The likelihood and the $-2\log\Lambda$ values are very similar for the three alternative models. Table 3.6 displays the statistics $-2\log\frac{L_{\text{alternative 3}}}{L_{\text{alternative 1}}}$ and $-2\log\frac{L_{\text{alternative 3}}}{L_{\text{alternative 1}}}$.

For each stock except for CLN alternative 1 is not significantly better than alternatives 2 or 3. This result is rather positive for following reasons:

Stock	$-2\log \dfrac{L_{\text{alternative2}}}{L_{\text{alternative1}}}$	$-2\log \dfrac{L_{\text{alternative3}}}{L_{\text{alternative1}}}$
ABB	1.54	2.58
ALUN	0.04	0.12
BALN	0.58	0.82
CIBN	0.14	0.16
CLN	20.86	20.86
CSGN	0.5	0.7
EMS	0.5	0.54
HOL	0.36	0.96
NESN	0.36	0.5
NOV	0.58	0.74
NOVN	0.66	5.58
ROG	0.38	0.44
RUKN	0.82	0.98
SGSI	0.96	1.5
SMH	0.42	1.78
SU/N	0.14	1.22
UBSN	0.06	0.3
ZURN	0.26	0.3

Table 3.6. $-2\log \dfrac{L_{\text{alternative 2/3}}}{L_{\text{alternative 1}}}$ statistics

- The sample estimates of the location and scale parameters can be used without significant loss of accuracy. This is particularly important in the M-dimensional case, where the location parameter is a M-dimensional vector and the scale parameter is $M \times M$-dimensional matrix. The estimation of M location, $\frac{(M+1)M}{2}$ scale and one number of degrees of freedom parameters can be still intractable for M large.
- The constraint that the number of degree of freedom must be an integer makes the simulation of the Student-t distribution easier. Indeed, in the one- and in the multidimensional case, a Student-t distributed variable Z can be obtained with:

$$Z = \frac{X}{\sqrt{\dfrac{S}{\nu}}} + \mu, \qquad (3.96)$$

where $X \sim N(0, \Sigma)$, $S \sim \chi^2_{(\nu)} = X_1^2 + X_2^2 + \cdots X_\nu^2$ with the X_i independent and $\sim \mathcal{N}(0, 1)$; X and S are independent.

The Multidimensional Case In this section the normal and Student-t models are extended to the multidimensional case and their descriptive power for the joint distribution of the 18 stocks is analysed. Note that both the multivariate normal and Student-t distributions belong to the class of elliptically contoured distributions (ECD), of which some characteristics are discussed in Sect. 2.1.2 (see also Sect. A.3.2). The class of ECD coincides with the class of multivariate distributions for which the covariance (or correlation) matrix is a sufficient measure of stochastic dependency, i.e. it completely describes the dependence between the risk factors (for a discussion see Wang (1997) and Embrechts et al. (1998)).

The density of a M-dimensional vector X with multivariate normal distribution is given by (see for instance Johnson (1987), p.50):

$$f(x|\mu, \sigma) = \frac{1}{(2\pi)^{\frac{M}{2}} \mid \Sigma \mid^{\frac{1}{2}}} \exp\left[-\frac{1}{2}(x - \mu)^{\top} \Sigma^{-1}(x - \mu)\right], \qquad (3.97)$$

where μ is the M-dimensional mean vector, Σ is the non-singular covariance matrix and $\mid \Sigma \mid$ is the determinant of Σ. Similar to the one-dimensional case, the maximum likelihood estimate of μ and Σ are the sample mean $\hat{\mu}$, respectively the sample covariance matrix S^2.

The density function of M-dimensional vector X with multivariate Student-t distribution is (see Johnson (1987), p.118):

$$f(x|m, H, \nu) = \frac{\Gamma\left(\frac{M+\nu}{2}\right)}{(\pi\nu)^{\frac{M}{2}} \Gamma\left(\frac{\nu}{2}\right)} \mid H \mid^{-\frac{1}{2}} \left[1 + \nu^{-1}(x - m)^{\top} H^{-1}(x - m)\right]^{-\frac{\nu+M}{2}},$$

$$(3.98)$$

where m is the M-dimensional location vector with $E(X) = m$, H is the scale matrix with $\text{cov}(X) = \frac{\nu-2}{\nu} H$, $\mid H \mid$ is the determinant of H and ν the number of degrees of freedom.

The log-likelihood function is given with:

$$l(m, H, \nu) = \sum_{i=1}^{n} \left[\log \Gamma\left(\frac{\nu + M}{2}\right) - \log \Gamma\left(\frac{\nu}{2}\right) - \frac{M}{2}\log(\pi\nu) + \log \mid H \mid^{-\frac{1}{2}}\right.$$

$$\left. - \frac{\nu + M}{2} \log\left[1 + \nu^{-1}(x_i - m)^{\top} H^{-1}(x_i - m)\right]\right].$$

$$(3.99)$$

In order to limit the number of parameters to optimise the sample mean and $\frac{n}{n-2}S^2$, where S^2 is the sample covariance matrix, are used as estimates of m respectively H and the log-likelihood function is optimised with respect to the number of degrees of freedom. The partial derivative of the log-likelihood function with respect to ν is:

$$\frac{\partial l}{\partial \nu} = \sum_{i=1}^{n} \left[\frac{1}{2} \psi \left[\frac{\nu + M}{2} \right] - \frac{1}{2} \psi \left[\frac{\nu}{2} \right] - \frac{M}{2\nu} - \frac{1}{2} \log \left[1 + \nu^{-1}(x_i - m)^\top H^{-1}(x_i - m) \right] \right.$$

$$\left. + \frac{\nu + M}{2} \frac{\left[(x_i - m)^\top H^{-1}(x_i - m) \right]}{\left[1 + \nu^{-1}(x_i - m)^\top H^{-1}(x_i - m) \right] \nu^2} \right].$$

$$(3.100)$$

Table 3.7 displays the maximum likelihood estimates and the statistics $-2 \log \Lambda$ for the normal model and two Student-t alternatives. In alternative 1 the number of degrees of freedom $\nu \in \mathbb{R}_+$. In alternative 2, $\nu \in \mathbb{N}$. As for the one-dimensional case, both Student-t alternatives outperform significantly the normal model and alternative 1 is not significantly better than alternative 2. The estimates of the numbers of degrees of freedom, 7.8 respectively 8, are however larger than almost all estimates of the one-dimensional case where the marginal distribution of the returns were considered separately. Clearly, as the dimension of the risk factors space increases, the diversification effect operates and the joint distribution tends to the normal one. The issue of fat tails is therefore more critical for portfolios exposed to a small number of risk factors. For well diversified portfolios the assumption of normal distribution is less unrealistic.

Model	log-likelihood	degrees of freedom	$-2\log \frac{L_{normal}}{L_{Student\,t}}$	$-2\log \frac{L_{alternative2}}{L_{alternative1}}$
Normal	51'993.10			
Student 1	52'069.41	7.81	1'562.27	0.28
Student 2	52'714.09	8.00	1'561.99	

Table 3.7. Maximum-likelihood statistics for the SMI portfolio

3.5.4 Impact of the Alternative Modelling

In Fig. 3.1 the empirical density of the daily log-returns on ABB is plotted with the densities of normal and Student-t models. The latter model fits the empirical data pretty well. The returns have too many small and large observations to be adequately described by the normal distribution. Table 3.8 shows percentile estimates of the return distributions under both distributional assumptions. The Student-t model produces larger estimates (in absolute terms) at levels lower than 3% and larger than 97%, and smaller ones at the intermediate levels. This has direct implication for risk management. The estimate of VaR numbers, for instance, will be more conservative under the alternative model only at very low levels. At the "standard" 5% level, the assumption of normality predicts larger VaR numbers than the Student-t model. Recall that extreme outcomes occur more often under the Student-t distribution than under the normal distribution, and therefore a small number of large shocks explain a large part of the total variance. The other shocks have a small dispersion around the mean. This is also consistent with other analysis of empirical returns (see for instance Hendricks (1996); Duffie and Pan (1997)). Duffie and Pan (1997) for example find the same patterns in the daily log-returns of the S&P 500 index.

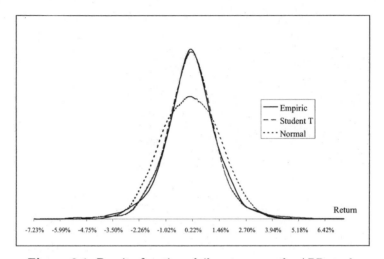

Figure 3.1. Density function: daily returns on the ABB stock

Table 3.9 shows the estimated α percentiles with ($\alpha = 0.5\%, 1\%, 3\%$ and 5%) of the profit and loss distributions (Value-at-Risk) for short option portfolios. The characteristics of the 18-dimensional portfolios are described in Sect. 1.6. The Value-at-Risk estimates are always higher under the Student-t distribution but particularly at the 0.5% and 1% levels. With increasing α both models produce estimates closer to another at least for PF_1^I and PF_2^I. These two portfolios consist of short calls, respectively short puts and exhibit a one-sided loss potential. By contrast, PF_3^I and PF_4^I, which consist of short straddles, respectively short butterfly spreads, have a two-sided loss potential, i.e. significantly market moves in either direction lead to losses. This explains why the VaR estimates under the Student-t distribution exceed significantly those predicted by the normal distribution even at the 5% level.

Percentile	Normal	Student-t	Difference %
0.50%	-3.23%	-4.10%	-26.89%
1.00%	-2.91%	-3.33%	-14.19%
1.50%	-2.71%	-2.92%	-7.62%
2.00%	-2.57%	-2.65%	-3.32%
2.50%	-2.45%	-2.45%	-0.17%
3.00%	-2.34%	-2.29%	2.28%
3.50%	-2.26%	-2.16%	4.27%
4.00%	-2.18%	-2.05%	5.94%
4.50%	-2.11%	-1.95%	7.37%
5.00%	-2.04%	-1.87%	8.61%
10.00%	-1.58%	-1.33%	16.01%
20.00%	-1.01%	-0.79%	22.22%
30.00%	-0.61%	-0.45%	25.69%
40.00%	-0.26%	-0.18%	30.21%
50.00%	0.06%	0.06%	0.00%
60.00%	0.39%	0.31%	20.52%
70.00%	0.73%	0.58%	21.35%
80.00%	1.14%	0.91%	19.80%
90.00%	1.70%	1.45%	14.84%
95.00%	2.17%	1.99%	8.12%
95.50%	2.23%	2.08%	6.96%
96.00%	2.30%	2.17%	5.62%
96.50%	2.38%	2.28%	4.05%
97.00%	2.47%	2.41%	2.17%
97.50%	2.57%	2.57%	-0.16%
98.00%	2.69%	2.77%	-3.17%
98.50%	2.84%	3.05%	-7.29%
99.00%	3.04%	3.45%	-13.61%
99.50%	3.36%	4.23%	-25.90%

Table 3.8. Percentile estimates: daily returns on the ABB stock

3.6 Continuous-Time Models

Although the returns of the risk factors are quoted at discrete intervals of time, it is often more convenient to model them as evolving in a continuous fashion through time. In particular the derivation of the Black-Scholes option-pricing formula requires the continuous adjustment of a portfolio replicating the option's pay-off exactly.

Let's assume that the prices of the risk factors follow a random walk with (this description bases on Campbell et al. (1997) pp.341-347):

$$w_k = w_{k-1} + \varepsilon_k, \qquad (3.101)$$

where w_0 is fixed, $k = 1, \cdots, n$ and ε_k is IID with values:

$$\varepsilon_k = \begin{cases} +\Delta \text{ with probability } \pi \\ -\Delta \text{ with probability } \pi' = 1 - \pi. \end{cases}$$

Underlying	Short		
$\alpha = 0.5\%$	$\bar{v}_{Normal}^{,\alpha}$	$\bar{v}_{Student}^{,\alpha}$	Diff %
PF_1^I	3618.16	4705.32	-23.10 %
PF_2^I	3358.69	3901.73	-13.92 %
PF_3^I	3062.99	4396.63	-30.33 %
PF_4^I	1046.91	1323.51	-20.90 %
$\alpha = 1\%$			
PF_1^I	3278.50	3914.65	-16.25 %
PF_2^I	3037.48	3228.55	-5.92 %
PF_3^I	2690.03	3565.00	-24.54 %
PF_4^I	961.35	1231.92	-21.96 %
$\alpha = 3\%$			
PF_1^I	2580.19	2833.75	-8.95 %
PF_2^I	2397.49	2440.80	-1.77 %
PF_3^I	2166.96	2614.51	-17.12 %
PF_4^I	823.20	1045.78	-21.28 %
$\alpha = 5\%$			
PF_1^I	2249.73	2292.72	-1.88 %
PF_2^I	2060.27	2092.76	-1.55 %
PF_3^I	1870.90	2244.82	-16.66 %
PF_4^I	744.91	944.92	-21.17 %

Table 3.9. Impact of alternative return models on the distribution $\bar{g}^I$

Consider a partition of the time interval $[0, T]$ into n periods of length $h = \frac{T}{n}$ and construct the continuous-time $\{w_n(t)\}$ process with:

$$w_n(t) = w_{[\frac{t}{h}]} = w_{[\frac{nt}{T}]}, \qquad (3.102)$$

where $[x]$ denotes the greatest integer $\leq x$. Although $\{w_n(t)\}$ is a continuous-time process it varies only at discrete points in time. $\{w_n(t)\}$ is therefore a discrete-state stochastic process which can be represented as a staircase function with discontinuities at the points $t = ih, (i = 1, 2 \cdots n)$. $w_n(T)$ can be rewritten as the sum of n independent variables w_i:

$$w_n(T) = w_1 + \cdots + w_n, \qquad (3.103)$$

where $E[w_i] = \pi\Delta - \pi'\Delta = \Delta(\pi - \pi')$, $E[w_i^2] = \pi\Delta^2 + \pi'\Delta^2 = \Delta^2(\pi + \pi') = \Delta^2$, and $\text{var}[w_i] = \Delta^2 - \Delta^2(\pi - \pi')^2 = 4\Delta^2\pi\pi'$.

This results in:

$$E[w_n(T)] = n(\pi - \pi')\Delta = \frac{T}{h}(\pi - \pi')\Delta,$$

$$\text{var}[w_n(T)] = 4n\pi\pi'\Delta^2 = 4\frac{T}{h}(\pi\pi')\Delta^2. \tag{3.104}$$

The construction of a well-defined and non-degenerate limiting process requires that the moments of $w_n(T)$ remain finite. To obtain the continuous-time version of the random walk, $\{w(t)\}$, Δ, π and π' must be adjusted so that:

$$E[w_n(T)] = \frac{T}{h}(\pi - \pi')\Delta \to \mu T,$$

$$\text{var}[w_n(T)] = 4\frac{T}{h}\pi\pi'\Delta^2 \to \sigma^2 T. \tag{3.105}$$

It can be easily seen that this is accomplished by setting:

$$\pi = \frac{1}{2}\left(1 + \frac{\mu\sqrt{h}}{\sigma}\right),$$

$$\pi' = \frac{1}{2}\left(1 - \frac{\mu\sqrt{h}}{\sigma}\right), \tag{3.106}$$

$$\Delta = \sigma\sqrt{h}.$$

It can be shown that $w_n(T)$ converges weakly to $\mathcal{N}(\mu T, \sigma^2 T)$, which implies that the distribution of $w_n(t)$ converges to $w(t)$ for all $t \in [0, T]$.

Thus, the process $\{w(t)\}$ has the following properties:

1. The increments have normal distribution, i.e.
$w(t_2) - w(t_1) \sim \mathcal{N}\left(\mu(t_2 - t_1), \sigma^2(t_2 - t_1)\right)$ for any t_1, t_2 with $0 \leq t_1 \leq t_2 \leq T$.

2. The increments $w(t_2) - w(t_1)$ are stationary and statistically independent of $w(t_4) - w(t_3)$ for any t_1, t_2, t_3, t_4 with $0 \leq t_1 \leq t_2 \leq t_3 \leq t_4 \leq T$.

3. The sample paths of $w(t)$ are continuous in t.

Setting $\mu = 0$ and $\sigma = 1$, $\{w(t)\}$ is referred to as a (standard) *Wiener process* and generally denoted by $W(t)$. As a consequence of properties 1 and 2 the sample paths of a Wiener process are not smooth and therefore the increment $W(t + h) - W(t)$ is not predictable given $W(t) - W(t - h)$. Observe that:

$$\text{var}\left[\frac{W(t + h) - W(t)}{h}\right] = \frac{1}{h^2}\sigma^2(t + h - t) = \frac{\sigma^2}{h} \tag{3.107}$$

and therefore the ratio $\to \infty$ as $h \to 0$. $W(t)$ is nowhere differentiable. Nevertheless, the infinitesimal increment of a Wiener process, i.e. $W(t + h) - W(t)$

as $h \to dt$ is commonly noted with $dW(t)$. Recall that the increments of a Wiener process are normally distributed with means zero and variance h. Following moments of $dW(t)$ may be calculated (see Shimko (1992), pp.4-5 for instance):

$$E[dW(t)] = \lim_{h \to dt} E[W(t+h) - W(t)] = 0$$

$$\mathrm{var}[dW(t)] = \lim_{h \to dt} E\left[(W(t+h) - W(t))^2\right] = dt$$

$$E[dW(t)dW(t)] = \lim_{h \to dt} E\left[(W(t+h) - W(t))^2\right] = dt$$

$$\mathrm{var}[dW(t)dW(t)] = \lim_{h \to dt} \left\{ E\left[(W(t+h) - W(t))^4\right] - h^2 \right\} = 3dt^2 - dt^2$$

$$\textit{since the fourth central moment of } \mathcal{N}(0,1) = 3$$

$$E[dW(t)dt] = \lim_{h \to dt} E\left[(W(t+h) - W(t))h\right] = 0$$

$$\mathrm{var}[dW(t)dt] = \lim_{h \to dt} E\left[(W(t+h) - W(t))^2 h^2\right] = dt^3.$$

$$(3.108)$$

As dt is infinitesimal, the terms dt^2 and dt^3 can be treated as essentially zero. $[dW(t)]^2$ and $dW(t)dt$ have therefore zero variance, i.e. are non-stochastic. $dW(t)$ is used as a basis for a whole family of continuous-time processes. For instance the stochastic differential equation for a variable following the continuous-time version of the random walk may be expressed as follows:

$$dw(t) = \mu dt + \sigma dW(t). \tag{3.109}$$

The process is referred to as *arithmetic Brownian Motion* with drift μ and volatility σ. Hence, the increment $dw(t)$ consists of a deterministic drift term μdt plus an additive Gaussian noise term $\sigma dW(t)$ and is therefore normally distributed. Since the support of the normal distribution is $\mathbb{R}$, the arithmetic Brownian motion allows a positive probability of negative w. This can be avoided by modelling $\{w(t)\}$ as a *geometric Brownian motion* :

$$dw(t) = \mu w(t)dt + \sigma w(t)dW(t). \tag{3.110}$$

If $w(t)$ obeys (3.110) then the proportional changes $\frac{dw}{w}$ are independent and identically normal distributed. Therefore the expected return is independent of the level of w, which makes sense. In the Black-Scholes framework the underlying price process is modelled as geometric Brownian motion. The generalisation of Brownian motions is referred to as *Itô processes* and are defined as :

$$dw(t) = a\big(w(t), t\big)dt + b\big(w(t), t\big)dW(t), \tag{3.111}$$

or, after dropping arguments:

$$dw = adt + bdW. \tag{3.112}$$

For stochastic variables described by a stochastic differential equation of Itô's type, the *Itô's Lemma* provides a stochastic version of the chain rule in that it relates the small change in a function f of one or several random variables to the small change(s) in the variable(s). A heuristic derivation of Itô's Lemma is given in Hull (2000), pp.235-236. Consider a smooth function f of two variables w and t. The bivariate Taylor series expansion of Δf is given with:

$$\Delta f = \frac{\partial f}{\partial w}\Delta w + \frac{\partial f}{\partial t}\Delta t + \frac{1}{2}\frac{\partial^2 f}{\partial w^2}(\Delta w)^2 + \frac{1}{2}\frac{\partial^2 f}{\partial t^2}(\Delta t)^2 + \frac{1}{2}\frac{\partial^2 f}{\partial w \partial t}(\Delta w)(\Delta t) + R,$$

(3.113)

where R is the remainder.

For w and t deterministic and $\Delta w \to dw \to 0$ and $\Delta t \to dt \to 0$, the second-order terms vanish so that:

$$df = \frac{\partial f}{\partial w}dw + \frac{\partial f}{\partial t}dt.$$

(3.114)

Suppose now that w is a stochastic variable following an Itô process. As shown above $[dW]^2$ has expectation dt and zero variance. It follows that:

$$\lim_{\Delta t \to 0}(\Delta w)^2 = b^2 dt.$$

(3.115)

Using this result df can be characterised by:

$$df = \frac{\partial f}{\partial w}dw + \frac{\partial f}{\partial t}dt + \frac{1}{2}\frac{\partial^2 f}{\partial w^2}b^2 dt,$$

(3.116)

and may be rewritten as follows by substituting dw with $a\,dt + b\,dW$:

$$df = \left(\frac{\partial f}{\partial w}a + \frac{\partial f}{\partial t} + \frac{1}{2}\frac{\partial^2 f}{\partial w^2}b^2\right)dt + \frac{\partial f}{\partial w}b\,dW.$$

(3.117)

Assuming now a geometric Brownian Motion (3.117) becomes:

$$df = \left(\frac{\partial f}{\partial w}\mu w + \frac{\partial f}{\partial t} + \frac{1}{2}\frac{\partial^2 f}{\partial w^2}\sigma^2 w^2\right)dt + \frac{\partial f}{\partial w}\sigma w\,dW,$$

(3.118)

and considering the function $f = \log(w)$ with following partial derivatives:

$$\frac{\partial f}{\partial w} = \frac{1}{w}, \quad \frac{\partial f}{\partial t} = 0, \quad \frac{\partial^2 f}{\partial w^2} = -\frac{1}{w^2},$$

(3.119)

it is obvious that logarithmic price changes:

$$df = \left(\mu - \frac{\sigma^2}{2}\right)dt + \sigma\,dW$$

(3.120)

are normally distributed with mean $\mu - \frac{\sigma^2}{2}$ and variance σ^2. Thus, the Geometric Brownian Motion is the continuous-time equivalent of the generic model.

3.7 Summary

Daily log-returns on risk factors are not independent and normally distributed as often assumed in finance theory and modelled in option pricing models (see Sect. 3.6). The well-documented evidence of excess kurtosis of the unconditional distributions and autocorrelation of the squared log-returns has been verified for the Swiss stock market. Both leptokurtosis and volatility clustering can be explained by changes in the conditional volatility. Of all models presented in Sect. 3.4, the GARCH and the stochastic volatility models have the most appealing characteristics since their unconditional distributions exhibit fat tails and both models can account for volatility clustering. However, it is recognised that both approaches, at least in their generic formulation with normally distributed innovations, do not adequately capture the excess kurtosis of the log-return data. On the other hand, the Student-t model, which can not account for volatility clustering, has shown good empirical results with respect to leptokurtosis, is parsimonious, and can easily be extended to the multivariate case. It has therefore been chosen as an alternative to the generic (normal) model in Sect. 3.5 and has provided a significantly better fit to the data than the generic normal. Thus, the empirical evidence strongly supports the Student-t model over the generic model. As shown in Sect. 3.5.4, the distributional form of returns has important implications for the estimation of risk measures. Different assumptions concerning the distribution of the risk factors may lead to substantial different risk measure estimates.

These results have an important implication for the measurement of market risks: methods used to approximate the portfolio distribution should be versatile enough to account for the no-normality of the risk factor distribution.

4. Valuation of Financial Instruments

The value of portfolio g at time T, denoted by $g[w(T)]$, is given with:

$$g[w(T)] = \sum_{i=1}^{I} \theta_i g_i[w(T)], \qquad (4.1)$$

where $g_i[w(T)]$ denotes the value of the ith investment and θ_i stands for the quantity of the ith *investment (asset, position)* held at time 0. The long and short positions of each instruments have been aggregated so that every θ_i represents a net investment. Clearly, the value of the portfolio depends on the level of the risk factors at time T and is therefore stochastic. Moreover, the modelling of the portfolio distribution F_g requires the valuation of the I financial instruments of the portfolio as functions of the risk factor values. In theory each asset could be modelled as a risk factor. For well diversified portfolios the procedure would require the distribution information of a large set of size I of instruments. Considering the idealised assumption that the I positions are jointly normal distributed, the computation of F_g would demand the estimation of I mean and $\frac{I(I+1)}{2}$ variance-covariance parameters. The parameter estimation for new, respectively illiquid instruments, would be very inaccurate, respectively impossible. Moreover, the modelling of instruments like options with non-linear pay-offs and therefore non-symmetric distributions would require the estimation of additional parameters since their distributions can not be completely described in terms of means and variances.

A better way of proceeding consists in decomposing each instrument of the portfolio of interest into basic instruments called *building blocks*. Here the building blocks are identified as *equities, bonds, futures (forwards)*, and *options*. Each of these building blocks is in turn a function of a limited number of risk factors. A typical set of risk factors may include the equity and commodity prices, currency exchange rates and interest rates on zero coupon bonds for all currencies of interest. All building blocks can be linked together in order to create more complex assets. For example a generic swap can either be modelled as positions in bonds, futures or options.

The main focus of this chapter is the valuation of the building blocks which may be categorised under *cash instruments* (equities and bonds) and *derivatives instruments* (forward and options). Cash instruments are traded in cash or spot markets where transactions are completed, i.e. where the ownership of the instruments is transferred immediately. By contrast derivative instruments, derivatives for short, are financial contracts whose value is related to the performance (i.e. price or fluctuation) of some underlying variable(s).

The next section discusses the principles of asset valuation and introduces the notion of pricing by *arbitrage* which is fundamental to the valuation of *contingent claims*. Sections 4.2 to 4.4 deal with the pricing of the single building blocks. Finally, Sect. 4.5 presents analytical approximations of the price functions which may be used when pricing equations do not have a closed-form solution.

4.1 Principles of Valuation

A financial asset represents a claim to some future benefit, generally to future cash flow(s). For valuation purposes a security is fully described by its payment schedule, i.e. by the timing and magnitude of its future cash flows. The price of an asset of interest at some time t may be obtained by discounting the cash flows to the time t. The idea that capital is simply future income discounted dates back to Fisher (1930) and is universally used in the finance practice. In a certain world, i.e. in a world where all future cash flows are known for sure, the valuation of all financial assets is straightforward and obtained by discounting the cash flows at the prevailing risk-free rate, which represents the reward for time.

Under uncertainty the discount function incorporates not only a reward for time but also for the risk of holding instruments with uncertain cash flows. Two types of valuations under uncertainty can be distinguished: the *absolute valuation method* and the *relative valuation method*. The *absolute valuation method* attempts to explain the price of all financial assets in terms of some fundamental values by modelling the preferences of market participants for consumption over time and possible states of the economy. A very well known result of the absolute valuation framework is the general equilibrium relationship for asset returns of the single period *Capital Asset Pricing Model (CAPM)*:

$$E[R_i(t, t+1)] = R_f(t, t+1) + \beta_i(t, t+1)E[R_M(t, t+1) - R_f(t, t+1)], \tag{4.2}$$

where $R_i(t, t+1)$ is the return of the ith asset over the period $(t, t+1)$, $R_M(t, t+1)$ denotes the return over the same period of a portfolio containing all risky assets, and $R_f(t, t+1)$ stands for the return of a riskless asset. $\beta_i(t, t+1)$ represents the quantity of *systematic risk* of asset i and yields:

$$\beta_i(t, t+1) = \frac{\text{cov}[R_M(t, t+1), R_i(t, t+1)]}{\text{var}[R_M(t, t+1)]}. \qquad (4.3)$$

Hence, the systematic risk is the part of the asset variance which is market-related. In the CAPM framework investors are rewarded only for bearing systematic risk. All non-systematic risk can be eliminated through diversification. The CAPM has been developed independently by Sharpe (1964) and Lintner (1965) and generalised by Mossin (1966).

Assuming that (4.2) holds, the price of an arbitrary asset at time t, $g_i(t)$, is therefore given with:

$$g_i(t) = \frac{E[g_i(t+1)] + E[d_i(t+1)]}{1 + E[R_i(t, t+1)]}, \qquad (4.4)$$

where $d_i(t+1)$ denotes eventual cash incomes such as coupons or dividends. Hence, the CAPM equilibrium price of an asset is equal to the expectation of its future value and cash future income discounted at a rate which incorporates a risk premium. The CAPM assumes either that all assets have elliptically symmetric distribution or that the investors have quadratic utility functions. As discussed in Sect. 2.1.2, there are strong assumptions, in particular for assets with non-symmetric pay-offs. A single period alternative to the CAPM is the *Arbitrage Pricing Theory* (APT) developed by Ross (1976). the APT makes weak assumptions on the behaviour of the market participants (non-satiation) but places restrictions on asset returns (and therefore on asset prices) which are represented as a linear combination of various common factors. The APT does not specifically identify these factors.

Note that financial assets are generally multiperiod instruments and according to (4.4) the valuation of the asset g_i requires an estimation of the cash flows and of the value of the asset at time $t+1$. Using (4.4) for period $(t+1, t+2)$:

$$g_i(t+1) = \frac{E[g_i(t+2)] + E[d_i(t+2)]}{1 + E[R_i(t+1, t+2)]}, \qquad (4.5)$$

and assuming that $E[g_i(t+1)]$ in (4.4) is equal to $g_i(t+1)$ in (4.5), (4.5) may be rewritten:

$$g_i(t) = \frac{E[d_i(t+1)]}{1 + E[R_i(t, t+1)]} + \frac{E[g_i(t+2)] + E[d_i(t+2)]}{(1 + E[R_i(t, t+1)])(1 + E[R_i(t+1, t+2)])}, \qquad (4.6)$$

Therefore by using formula (4.4) recursively, the price of an asset with maturity $t+u$ is given with:

$$g_i(t) = \sum_{k=t}^{t+u} \frac{E[d_i(k+1)]}{\prod_{j=t}^{k}(1 + E[R_i(j, j+1)])}, \tag{4.7}$$

where it is assumed in any period that the expectations are fulfilled ex post. This assumption is referred to as the *rational expectation hypothesis* and is generally implicitly made in the various discrete multiperiod or continuous-time extensions of the CAPM, which relax some of the assumptions of the single-period CAPM. A detailed overview of discrete time models can be found in Ingersoll (1987). Continuous-time models are presented extensively in Merton (1992).

The second approach to valuation, the *relative valuation* method, has a more modest scope since it explains the price of some assets, called *contingent claims*, in terms of other given and observable asset prices. The relative valuation method makes no assumption regarding the probability distribution of the observable asset prices and places very weak restrictions on the behaviour of the market participants, namely non-satiation. This relative pricing approach is based on the concept of *arbitrage* and is therefore also referred to as *valuation by arbitrage*. The valuation by arbitrage is very useful for the pricing of derivatives and is discussed in the next section. Intuitively, an arbitrage opportunity arises if an investment in a portfolio of assets guarantees a riskless profit higher than the risk-free return. An economy where investors are non-satiated, i.e. prefer more to less, will rule out arbitrage opportunities regardless of investors' risk tolerances. One consequence of the non-arbitrage condition is the *single price law of markets* which states that two investments with the same pay-off in every state of the world must have the same current value. Another fundamental result of the arbitrage theory is that for a (complete) market without arbitrage opportunity, there exists a unique probability measure Q *equivalent* to P, i.e. Q and P assigns a positive probability to the same events, under which discounted prices are martingales to be defined. As a result, the current price of any contingent claim is equal to its discounted future pay-offs under Q. This result is of central importance for the pricing of derivatives and is discussed in the following section.

4.1.1 Valuation by Arbitrage

The valuation by arbitrage is widely used for the pricing of derivatives since the pioneering works of Harrison and Kreps (1979), and Harrison and Pliska (1981). Section 4.1.1 discusses the approach for a single period economy with finite sample space Ω. The model is extended in Sect. 4.1.1 to a discrete-time multiperiod setting. Finally, Sect. 4.1.1 discusses some particularities of the modelling in continuous-time.

Single-Period, Finite State Model

Following the conceptual layouts of Bingham and Kiesel (1998), pp.20-30, and particulary of Pliska (1997), pp.1-28, let's consider a single period model with time indices $t = 0$, the initial time, and $t = T$, the terminal time and a finite probability space, i.e. a space with finite cardinality, $\Omega = \{\omega_1, \cdots, \omega_K\}$. P is defined as a probability on the events of Ω, $B = \{B(t) : t = 0, T\}$ is a special asset called the *bank account process* with $B(0) = 1$ and $B(T) > 0$ (and generally ≥ 1) for all $\omega \in \Omega$. $B(T) - 1 = r$ may therefore be interpreted as the interest rate of the bank account. $B(T)$ and hence r need not to be deterministic. $S = \{S(t) : t = 0, T\}$ is a I-dimensional *price process* where $S_i(t)$ denotes the price of security i ($i = 1, \cdots, I$) at time t. A *trading strategy* $\theta = (\theta_0, \theta_1, \cdots, \theta_I)$ represents the portfolio held at time $t = 0$ by an agent. θ_0 denotes the amount of money invested in the bank account and θ_i ($i = 1, \cdots, I$) is the number of shares of security i held at time $t = 0$. Short holdings are allowed. During the period $]0, T]$ re-balancing is not permitted. The process $V = \{V(t) : t = 0, T\}$, which describes the value of the portfolio corresponding to the trading strategy θ, is referred to as the *value process* and is obviously:

$$V(t) = \theta_0 B(t) + \sum_{i=1}^{I} \theta_i S_i(t). \tag{4.8}$$

The *profit and loss* of the trading strategy is called the gain process G is given with:

$$G = \theta_0\, r + \sum_{i=1}^{I} \theta_i \Delta S_i, \tag{4.9}$$

where $\Delta S_i = S_i(T) - S_i(0)$. The profit and loss G may be represented by a vector in $\mathbb{R}^K$. Each entry $G(\omega_k)$, with $k = 1, \cdots, K$, represents the value of G in a given state ω_k.

In order to simplify the development, it is advantageous to change the *numéraire*, to express the price process as a function of another unit of measure. If the bank account is chosen as numéraire, the *discounted price process*, is defined as follows:

$$S_i^*(t) = S_i(t)/B(t). \tag{4.10}$$

Note that $S_i^*(0) = S_i(0)$ since $B(0) = 1$. Similarly, the *discounted value process* $V^* = \{V^*(t) : t = 0, T\}$ yields:

$$V^*(t) = \theta_0 + \sum_{i=1}^{I} \theta_i S_i^*(t), \tag{4.11}$$

and the *discounted gain process* and is given with:

$$G^* = \sum_{i=1}^{I} \theta_i \Delta S_i^*, \qquad (4.12)$$

where $\Delta S_i^* = S_i^*(T) - S_i^*(0)$. It is worth noting that any asset with strictly positive pay-offs could be chosen as numéraire.

A trading strategy $\hat{\theta}$, with value process $\hat{V}$, is said to *dominate* another trading strategy $\tilde{\theta}$, with value process $\tilde{V}$, if:

$$\hat{V}(0) = \tilde{V}(0) \text{ and } \hat{V}(T, \omega) > \tilde{V}(T, \omega) \ \forall \omega \in \Omega, \qquad (4.13)$$

where $V(T, \omega)$ denotes the pay-off of V at time T given outcome $\omega \in \Omega$. By setting $\theta = \hat{\theta} - \tilde{\theta}$ the value process V of θ is given with:

$$V(0) = 0 \text{ and } V(T, \omega) > 0 \ \forall \omega \in \Omega, \qquad (4.14)$$

and represents a portfolio with no initial cost and strictly positive value at time T for all $\omega \in \Omega$.

It is easy to verify that the existence of a dominant strategy, as defined in (4.13), allows to build another strategy $\check{\theta}$ so that:

$$\check{V}(0) < 0 \text{ and } \check{V}(T, \omega) \geq 0 \ \forall \omega \in \Omega, \qquad (4.15)$$

which represents a portfolio with negative commitment and non-negative value at time T for all $\omega \in \Omega$.

Hence, the dominant strategy offers larger, or at least equal, pay-offs in every possible future state of the world for the same initial cost. Clearly, the existence of such a strategy is not satisfactory and should not be allowed by the model. All (non-satiated) investors would hold the dominated portfolio and sell the dominated one.

Note that the notion of dominance considered here, also called absolute dominance in the sequel, is even stronger than first-order stochastic dominance (see Sect. 2.1.2), which will be referred to as weak dominance in the following. A (first-order) stochastically dominant strategy (or portfolio) needs not to outperform the dominated strategy in every state, but its probability of exceeding any given value is higher than that of a stochastically dominated strategy.

Mathematically, the absolute dominance of strategy $\hat{\theta}$ over strategy $\tilde{\theta}$ implies that (Ingersoll (1987), p.123):

$$\tilde{V}(T) = \hat{V}(T) + \eta, \qquad \tilde{V}(0) = \hat{V}(0), \qquad (4.16)$$

where η is a non-positive random variable, i.e. a random variable whose outcome is non-positive for every state of the world and negative for at least one ω. In the event of weak dominance of $\hat{\theta}$ over $\tilde{\theta}$, (4.16) becomes:

$$\tilde{V}(T) \stackrel{d}{=} \hat{V}(T) + \eta, \qquad \tilde{V}(0) = \hat{V}(0), \tag{4.17}$$

where $\stackrel{d}{=}$ means equal in distribution (see Sect. A.2).

A further consequence of the existence of an absolutely dominant strategy is that the pricing in the economy is not logical since strategy $\hat{\theta}$, which offers a greater pay-off than $\tilde{\theta}$ at time T for all states of the world, has the same initial value as $\tilde{\theta}$. Obviously, the existence of a *linear pricing measure* $\pi = (\pi(\omega_1), \cdots, \pi(\omega_K))$ with $\pi(\omega_k) \geq 0, k = 1, \cdots, K$:

$$V^*(0) = \sum_{k=1}^{K} \pi(\omega_k) V^*(T, \omega_k) = \sum_{k=1}^{K} \pi(\omega_k) V(T, \omega_k)/B(T, \omega_k), \tag{4.18}$$

ensures a consistent pricing since each strategy with strictly greater (smaller) pay-off at time T has a higher (lower) initial price. Moreover, linear programming arguments show that the existence of a linear pricing measure rules out the existence of dominant strategy.

Let Z be the $(N+1)$-columns vector of the discounted initial prices:

$$Z = \begin{pmatrix} S_1^*(0) \\ \vdots \\ S_I^*(0) \\ 1 \end{pmatrix},$$

and $\mathcal{Z}$ the $(N+1) \times K$ matrix of the discounted terminal prices:

$$\mathcal{Z} = \begin{pmatrix} S_1^*(T, \omega_1) & \cdots & S_1^*(T, \omega_K) \\ \vdots & \cdots & \vdots \\ S_I^*(T, \omega_1) & \cdots & S_I^*(T, \omega_K) \\ 1 & \cdots & 1 \end{pmatrix}.$$

Clearly, if a linear pricing measure exists, the following linear program:

$$\begin{aligned} \text{maximise} \quad & (0, \cdots, 0) \cdot \pi, \\ \text{subject to} \quad & \mathcal{Z}\pi = Z, \\ & \pi \geq 0, \end{aligned} \tag{4.19}$$

has a solution.

From the *theorem of duality* (see for instance Ingersoll (1987), pp.10-11), a finite objective maximum for the primal (4.19) guarantees that the dual:

$$\text{minimise} \quad hZ,$$
$$\text{subject to} \quad h\mathcal{Z} \geq 0, \tag{4.20}$$

is feasible. Moreover, the objective values of programs (4.19) and (4.20) must be equal and from (4.19) must yield zero. Since h may be interpreted as a trading strategy, hZ is therefore equal to the initial discounted value $V^*(0)$ of the trading strategy h (see equation(4.11)). As a result, the existence a linear pricing measure ensures that the initial discounted value is (at least) zero whereas the terminal discounted value is greater than or equals zero. Conversely, if there is no dominant strategy, the solution of (4.20) is trivially $h = 0$ and from the theorem of duality there exists a solution π for (4.19), which ensures that a linear pricing measure exists. In other words, dominant trading strategies are ruled out if, and only if, there exists a linear pricing measure. Note that the absence of dominant trading strategy implies that the *law of one price* holds, i.e. $\tilde{V}(T,\omega) = \hat{V}(T,\omega)$ for all $\omega \in \Omega$ implies that $\tilde{V}(0) = \hat{V}(0)$. The converse, however, is not true (for a numerical example see Pliska (1997), p.8).

In the event of an existing (non-negative) linear pricing measure (4.11) may be rewritten for $t = 0$:

$$\theta_0 + \sum_{i=1}^{I} \theta_i S_i^*(0) = \sum_{k=1}^{K} \pi(\omega_k) \left[\theta_0 + \sum_{i=1}^{I} \theta_i S_i^*(T, w_k) \right]. \tag{4.21}$$

By setting $\theta_i = 0$, $i > 0$, in (4.21), it is easy to verify that $\pi(\omega_1) + \cdots + \pi(\omega_K) = 1$ and therefore π has the properties of a probability measure on Ω (see A.1). Moreover, for a holding strategy θ so that $\theta_i = 0$ for all i except an arbitrary $j \in \{1, \cdots, I\}$, (4.21) becomes:

$$S_j^*(0) = \sum_{k=1}^{K} \pi(\omega_k) S_j^*(T, \omega_k), \qquad j = 1, \cdots, I \tag{4.22}$$

and therefore the initial price of asset S_j equals the expectation of the discounted price at time T under the probability measure π.

An *arbitrage opportunity* is defined as a trading strategy θ, whose value process V exhibits following properties:

$$V(0) = 0,$$
$$V(T) \geq 0, \tag{4.23}$$
$$E_P[V(T)] > 0,$$

where $E_P[.]$ denotes the expectation operator under probability P.

As the initial cost of the trading strategy described in (4.23) is zero and its terminal value is non-negative, an arbitrage opportunity represents a riskless possibility of profit. The concept of arbitrage opportunity is weaker than the concept of dominant strategy since it does not ensure a strictly positive terminal value, i.e. $V(T)$ may be zero for some $\omega \in \Omega$. As a result, a dominant trading strategy implies an arbitrage opportunity while the converse is not necessarily true. Clearly, if arbitrage opportunities exist, non-satiated market participants will take unbounded positions in these arbitrage opportunities. This is not consistent with market equilibrium.

Since $B(t)$ is positive for all t definition (4.23) may be rewritten as (see also (4.10)):

$$
\begin{aligned}
V^*(0) &= 0, \\
V^*(T) &\geq 0, \\
E_P[V^*(T)] &> 0,
\end{aligned}
\tag{4.24}
$$

which is equivalent to (see equations (4.11) and (4.12)):

$$
\begin{aligned}
G^* &\geq 0, \\
E_P[G^*] &> 0.
\end{aligned}
\tag{4.25}
$$

As it will be shown in the sequel, there is no arbitrage opportunity if, and only if, there exists a risk-neutral probability measure Q defined as follows:

1. $Q(\omega) > 0 \quad \forall \omega \in \Omega$
2. $E_Q[\Delta S_i^*] = 0 \quad i \in \{1, \cdots, I\}$,

where $E_Q[\cdot]$ denotes the expectation operator under probability Q. A risk-neutral measure is therefore a linear pricing measure giving a strictly positive probability to every state in Ω. From definition (4.25) it follows that the absence of arbitrage opportunity implies both sets

$$
W = \{X \in \mathbb{R}^K : X = G^* \quad \text{for some } \theta\},
\tag{4.26}
$$

and

$$
\mathbb{R}_+^K = \{X \in \mathbb{R}^K : X_k \geq 0 \quad \forall k \ (0 \leq k \leq K) \quad \exists k \text{ so that } X_k > 0\}, \tag{4.27}
$$

having no common points since all elements of $\mathbb{R}_+^K$ have exclusively non-negative (and at least one strictly positive) components and therefore a positive expected value.

Define with $W^\perp$ the orthogonal complement of W:

$$
W^\perp = \{Y \in \mathbb{R}^K : X \cdot Y = 0 \quad \forall X \in W\},
\tag{4.28}
$$

where $X \cdot Y$ denotes the inner product of X and Y, and with A^+ following convex and closed subset of $\mathbb{R}_+^K$:

$$A^+ = \{X \in \mathbb{R}_+^K : E[X] \geq 1\}, \tag{4.29}$$

In the absence of arbitrage opportunity, $A^+ \subset \mathbb{R}_+^K$ and W are disjoint. From the *separating hyperplane theorem* (see Duffie (1996), pp.275-277 for a formal discussion) it can be shown that:

$$\exists Y \in W^\perp \text{ so that } X \cdot Y > 0 \quad \forall X \in A^+. \tag{4.30}$$

Note that for each $k = 1, \cdots, K$ there is a vector $X \in A^+$ so that its kth entry is positive and all the others equal zero. From (4.30) it results that Y must be strictly positive and by defining:

$$Q(\omega_k) = \frac{Y(\omega_k)}{Y(\omega_1) + \cdots + Y(\omega_K)}, \tag{4.31}$$

it is obvious that $Q(\omega) \in W^\perp$ with $Q(\omega) > 0$ for all ω and $\sum_{i=1}^K Q(\omega_i) = 1$. From the definition of the discounted gain process (4.12):

$$G^* = \sum_{i=1}^I \theta_i \Delta S_i^*.$$

it is straightforward that there exists a trading strategy for each i so that:

$$G^* = \Delta S_i^*. \tag{4.32}$$

As a result, ΔS_i^* is an element of W for all i and it results that:

$$E_Q[\Delta S_i^*] = 0 \quad i \in \{1, \cdots, I\} \tag{4.33}$$

Hence $Q(\omega)$ may be interpreted as a risk-neutral probability and therefore the absence of arbitrage opportunity implies the existence of a risk-neutral probability measure.

To prove the converse recall the definition of a discounted gain process (4.12):

$$G^* = \sum_{i=1}^I \theta_i \Delta S_i^*.$$

If a risk-neutral probability measure Q exists then following holds:

$$E_Q[G^*] = E_Q\left[\sum_{i=1}^I \theta_i \Delta S_i^*\right] = \sum_{i=1}^I \theta_i E_Q[\Delta S_i^*] = 0 \tag{4.34}$$

In other words there exists a strictly positive vector Q so that:

$$Q^\top G^* = 0, \tag{4.35}$$

and therefore it is not possible that both conditions of definition (4.25) are simultaneously satisfied. As a result, the model is free of arbitrage opportunities if, and only if, there exists a risk-neutral probability measure.

A *contingent claim* $g(T)$ is a non-negative random variable which represents some pay-off at the terminal date T. $g(T)$ is said to be *attainable* (or *marketable*) if there exists a trading strategy θ, so that:

$$g(T) = V(T). \tag{4.36}$$

The trading strategy is said to generate the contingent claim $g(T)$. In the absence of arbitrage opportunities, the law of one price requires that:

$$g(0) = V(0), \tag{4.37}$$

and therefore from (4.8):

$$g(0) = \theta_0 B(t) + \sum_{i=1}^{I} \theta_i S_i(t). \tag{4.38}$$

Since the absence of arbitrage opportunities implies the existence of a risk-neutral probability measure, it follows that (see equations (4.11), (4.12) and (4.34)):

$$
\begin{aligned}
g(0) &= V(0) - V^*(0), \\
&= E_Q[V^*(T) - G^*], \\
&= E_Q[V^*(T)] - \sum_{i=1}^{I} \theta_i E_Q[\Delta S_i^*], \\
&= E_Q[V^*(T)] - 0, \\
&= E_Q[V(T)/B(T)], \\
&= E_Q[g(T)/B(T)],
\end{aligned} \tag{4.39}
$$

and hence the initial value of a marketable contingent claim equals the expectation under Q of its discounted pay-off. If every contingent claim can be generated by a trading strategy, the market is referred to as *complete*. Let $\mathcal{G}$ $(T) \in \mathbb{R}^K$ be the vector of the possible pay-offs of an arbitrary contingent claim $g(T)$:

$$\mathcal{G}(T) = [g(T, w_1), \cdots, g(T, w_K)], \tag{4.40}$$

and $S(T) \in \mathbb{R}^{K \times (I+1)}$ the matrix of the terminal prices of the $I+1$ assets available:

$$S(T) = \begin{pmatrix} B(T,w_1) & S_1(T,\omega_1) & \cdots & S_I(T,\omega_1) \\ \vdots & \vdots & \cdots & \vdots \\ B(T,w_K) & S_1(T,\omega_K) & \cdots & S_I(T,\omega_K) \end{pmatrix}. \tag{4.41}$$

It is easy to verify that:

$$S(T)\theta = \mathcal{G}(T) \tag{4.42}$$

has a solution if, and only if, the matrix $S(T)$ has rank K. In other words, (4.42) has a solution and therefore the market is complete if the number of independent vectors in $S(T)$ is equal to the number of states of the world. Moreover, the market is complete if, and only if, there exists a unique risk-neutral probability measure (see Pliska (1997), pp.23-24 for a derivation).

Multiperiod Discrete-time Model

Let's consider a multiperiod model (see Bingham and Kiesel (1998), pp.83-102, and Pliska (1997), pp.72-100) with $T + 1$ discrete trading dates $t = 0, 1, \cdots, T$ and with finite sample space Ω equipped with a filtration $\mathbb{F} = \{\mathcal{F}(t) : t \in \mathcal{T}\}$. The positive one-dimensional bank account process $B = \{B(t) : t = 0, 1, \cdots, T\}$ and the I-dimensional price process $S = \{S(t) : t = 0, 1, \cdots, T\}$ are discrete and assumed to be *adapted* to the $\mathbb{F}$, i.e. the variables $S_i(t)$ with $i = 1, \cdots, I$ and $B(t)$ are measurable with respect to $\mathcal{F}(t)$ for every t (see Sects. 1.3.2 and A.2.5 for a more detailed discussion). Note that the *filtration*, which is a complete description of the information structure, is a non-decreasing sequence of σ-algebras since it is assumed that new information about the states of the world is coming in as time elapses and nothing is forgotten. As the sample space is finite the information structure is equivalently described by a sequence of *partitions* $\mathcal{P}(t)$ of Ω. Each partition is a collection of disjoints subsets of the sample space whose union is Ω. It is assumed, without loss of generality, that the filtration $\mathcal{F}(0)$ is the trivial algebra $\{\emptyset, \Omega\}$ and that the filtration $\mathcal{F}(T)$ corresponds to the power set 2^Ω. The corresponding partitions $\mathcal{P}(0)$, respectively $\mathcal{P}(T)$, are $\{\Omega\}$, respectively $\{\{w_1\}, \cdots, \{w_K\}\}$. In other words, no information is available at time $t = 0$ and individuals only know that the "true state of the world", ω_{true} is an element of Ω. As time progresses the true state of the world is gradually revealed to the market participants, i.e. the partition of Ω becomes increasingly finer and therefore the cardinality of the event containing ω_{true} decreases. At time $t = T$ the true state of the world is fully revealed.

The stochastic process describing the trading strategy is called *predictable* with respect to the filtration $\mathbb{F}$, i.e. is measurable with respect to $\mathcal{F}(t - 1)$. The value process $V = \{V(t); t = 0, 1, \cdots, T\}$ is given with:

$$V(t) = \begin{cases} \theta_0(1)B(0) + \sum_{i=1}^{I} \theta_i(1)S_i(0), & t = 0 \\ \theta_0(t)B(t) + \sum_{i=1}^{I} \theta_i(t)S_i(t), & t \geq 1 \end{cases} \tag{4.43}$$

and therefore $V(0)$ is the initial value of the portfolio while $V(t)$ is the portfolio value at time t *before* the transactions made at t. A trading strategy θ is referred to as *self-financing* trading strategy if:

$$V(t) = \theta_0(t+1)B(t) + \sum_{i=1}^{I} \theta_i(t+1)S_i(t), \qquad t = 1, \cdots T-1 \qquad (4.44)$$

i.e. a strategy which does not involve cash infusions into or cash withdrawals from the portfolio. Similarly to the single period setting, the discounted price processes $S_i^* = \{S_i^*(t); t = 0, 1, \cdots, T\}$ for $i = 1, \cdots, I$ are defined by:

$$S_i^*(t) = S_i(t)/B(t), \qquad (4.45)$$

the discounted value process $V^* = \{V^*(t); t = 0, 1, \cdots, T\}$ yields:

$$V^*(t) = \begin{cases} \theta_0(1)B(0) + \sum_{i=1}^{I} \theta_i(1)S_i^*(0), & t = 0 \\ \theta_0(1)B(t) + \sum_{i=1}^{I} \theta_i(t)S_i^*(t), & t \geq 1 \end{cases} \qquad (4.46)$$

and the discounted gain process $G^* = \{G^*(t); t = 1, 2, \cdots, T\}$ is given with:

$$G^*(t) = \sum_{i=1}^{I} \sum_{j=1}^{t} \theta_i(j)\Delta S_i^*(j), \qquad (4.47)$$

where $\Delta S_i^*(j) = S_i^*(j) - S_i^*(j-1)$.

It is easy to verify that θ is self-financing with respect to the value process if, and only if, θ is self-financing with respect to the discounted value process:

$$\theta(t)B(t) + \sum_{i=1}^{I} \theta_i(t)S_i(t) = \theta_0(t+1)B(t) + \sum_{i=1}^{I} \theta_i(t+1)S_i(t)$$

$$\Leftrightarrow \qquad (4.48)$$

$$\theta_0(t) + \sum_{i=1}^{I} \theta_i(t)\frac{S_i(t)}{B(t)} = \theta_0(t+1) + \sum_{i=1}^{I} \theta_i(t+1)\frac{S_i(t)}{B(t)}.$$

This property is referred to as *numéraire invariance*. Moreover, if θ is self-financing it implies for $t = 1, 2, \cdots, T$:

$$V^*(0) + G^*(t) = \theta_0(1) + \sum_{i=1}^{I} \theta_i(1)S_i^*(0) + \sum_{i=1}^{I}\sum_{j=1}^{t} \theta_i(j)\left(S_i^*(j) - S_i^*(j-1)\right)$$

from equations (4.46) and (4.47)

$$= \theta_0(1) + \sum_{i=1}^{I} \theta_i(1)S_i^*(0) + \sum_{i=1}^{I} \theta_i(t)S_i^*(t) + \sum_{i=1}^{I}\sum_{j=1}^{t-1} \theta_i(j)S_i^*(j)$$

$$- \sum_{i=1}^{I}\sum_{j=2}^{t} \theta_i(j)S_i^*(j-1) - \sum_{i=1}^{I} \theta_i(1)S_i^*(0)$$

$$= \theta_0(1) + \sum_{i=1}^{I} \theta_i(t)S_i^*(t) + \sum_{i=1}^{I}\sum_{j=1}^{t-1} \left[\theta_i(j) - \theta_i(j+1)\right]S_i^*(j)$$

$$= \theta_0(1) + \sum_{j=1}^{t-1} \left[\theta_0(j+1) - \theta_0(j)\right] + \sum_{i=1}^{I} \theta_i(t)S_i^*(t)$$

since θ is self-financing and therefore from (4.48):

$$\sum_{i=1}^{I} \left[\theta_i(j) - \theta_i(j+1)\right]S_i^*(j) = \theta_0(j+1) - \theta_0(j)$$

$$= \theta_0(t) + \sum_{i=1}^{I} \theta_i(t)S_i^*(t)$$

$$= V^*(t).$$

$$(4.49)$$

In the multiperiod model an *arbitrage opportunity* is a trading strategy θ with value process V satisfying:

$$\begin{aligned} &V(0) = 0 \\ &V(T) \geq 0, \\ &E_P[V(T)] > 0, \\ &\theta \text{ is self-financing.} \end{aligned} \qquad (4.50)$$

From (4.46) and from the numéraire invariance property of self-financing trading strategies it is obvious that a self-financing trading strategy constitutes an arbitrage opportunity if:

$$\begin{aligned} &V^*(0) = 0, \\ &V^*(T) \geq 0, \\ &E_P[V^*(T)] > 0. \end{aligned} \qquad (4.51)$$

or equivalently (see (4.47)):

$$\begin{aligned} &G^*(T) \geq 0, \\ &E_P[G^*(T)] > 0, \end{aligned} \qquad (4.52)$$

As it will be discussed in the sequel, arbitrage opportunities are ruled out if, and only if, there is a probability measure Q, so that:

1. Q is equivalent to P (see Sect. A.1), which implies that $Q(\omega) > 0$,
2. the price process S_i^* is a *martingale* under Q for all $i \in \{1, 2, \cdots I\}$, i.e.:

$$E_Q\left[S_i^*(t+s) \mid \mathcal{F}(t)\right] = S_i^*(t) \qquad t, s \geq 0. \tag{4.53}$$

The left-hand side of (4.53) represents the expectation of $S_i^*(t)$ under Q conditional on the σ-algebra $\mathcal{F}(t)$, i.e. on the information available at time t. A more formal definition of the conditional expectation and martingale as well as a list of their important properties can be found in Sect. A.2.4. Note that Q is then said to be a *risk-neutral probability measure* or *equivalent martingale measure*.

The existence of an equivalent martingale measure Q rules out arbitrage opportunities. Since θ is self-financing it implies that (see (4.49)):

$$V^*(t) = V^*(0) + G^*(t). \tag{4.54}$$

From equations (4.54) and (4.47) it is easy to verify that:

$$V^*(t+1) - V^*(t) = G^*(t+1) - G^*(t)$$
$$= \sum_{i=1}^{I} \theta(t+1)\left(S_i^*(t+1) - S_i^*(t)\right), \tag{4.55}$$

and therefore:

$$V^*(t) = V^*(0) + \sum_{i=1}^{I} \sum_{t=1}^{T} \theta(t)\left(S_i^*(t) - S_i^*(t-1)\right). \tag{4.56}$$

Thus, $V^*(t)$ is the *martingale transform* of the Q-martingale S^* by θ and is therefore a martingale under Q (see appendix A.2.6). Since the discounted price process is a martingale, it follows that $V^*(0) = E_Q[V^*(T)]$ and therefore self-financing trading strategies may not be arbitrage opportunities (see definition (4.51): if $V^*(T) \geq 0$ and $E_P[V^*(T)] > 0$ it implies that $E_Q[V^*(T)] > 0$ since Q and P are equivalent).

The converse, i.e. the fact that absence of arbitrage opportunities implies the existence of an equivalent martingale measure, can be shown by extending the one-period model. Let's denote with $A'(t)$ an arbitrary event, subset of the partition $\mathcal{P}(t)$. $A'(t)$ may be thought as the disjoint union of the events $A_l'(t+1) \in \mathcal{P}(t+1)$, with $l = 1, \cdots, L$ since the information structure is made up of a sequence of increasingly finer partitions. Furthermore, since the price and the discount processes are adapted it results that $S^*(t)$ is constant on $A(t)$ and $S^*(t+1)$ is constant on each $A_l'(t+1)$ (see A.2.5). In other words, given that event $A(t)$ occurs at time t, the events $A'_l(t+1)$ correspond to the possible events which may prevail at time $t+1$. Hence, the multi-period setting may be modelled as a collection of single-period markets. Let's assume that there exists an arbitrage opportunity $\tilde{\theta}$ conditional on the event $A'(t)$, i.e. $\sum_{i=1}^{I} \tilde{\theta}_i \Delta S_i^*(t+1) \geq 0$ and $E[\sum_{i=1}^{I} \tilde{\theta}_i \Delta S_i^*(t+1)] > 0$ given that $A'(t)$ occurs at time t. It is easy to verify that a trading strategy θ with:

$$\theta_i(u,\omega) = \begin{cases} 0 & u \leq t, \\ 0 & u = t+1, \omega_{\text{true}} \in \Omega \setminus A'(t), \\ \tilde{\theta}_i & u = t+1, \omega_{\text{true}} \in A'(t), i = 1,\cdots,I, \\ -\sum_{i=1}^{I} \tilde{\theta}_i S_i^*(t) & u = t+1, \omega_{\text{true}} \in A'(t), i = 0 \\ 0 & u > t+1, \omega_{\text{true}} \in A'(t), i = 1,\cdots,I, \\ -\sum_{i=1}^{I} \tilde{\theta}_i & u > t+1, \omega_{\text{true}} \in A'(t), i = 0 \end{cases} \quad (4.57)$$

and where ω_{true} denotes the true state of the world, is self-financing and constitutes an arbitrage opportunity for the multiperiod model. Hence, if the multiperiod model does not accept arbitrage opportunities it results that single period arbitrage opportunities (conditional on an arbitrary event) are not allowed. The no-arbitrage condition implies therefore that there exists a probability measure $Q_{(t,A(t))}$ for each $A(t) \in \mathcal{P}(t)$ so that:

$$E_{Q_{(t,A(t))}}[S_i^*(t+1)] = S_i^*(t) \quad \forall i \quad \forall t < T. \quad (4.58)$$

$Q_{(t,A(t))}$ is clearly a risk-neutral probability measure conditional on the event $A(t)$ and gives therefore a strictly positive mass to each event $A'_l(t+1)$. Let's define a *path* (of the discounted price process) as the mapping $t \to S_n^*(t,\omega_k)$ for a given $\omega_k \in \Omega$ and let's denote by Q a measure which associates to each single event ω_k a weighting $Q(\omega)$ equal to the product of the risk-neutral probabilities along the corresponding path:

$$Q(w_k) = \prod_{t=1}^{T-1} Q_{(t,A_k(t))} \quad \text{where } A'_k(0) \supset A'_k(1) \supset \cdots \supset w_k. \quad (4.59)$$

Clearly, $Q(w_k) > 0$ since the conditional probabilities are strictly positive. Since $Q(w_k)$ is the product of conditional probabilities it is obvious that:

$$\sum_{k=1}^{K} Q(w_k) = 1, \quad (4.60)$$

and therefore $Q(w_k)$ is an equivalent probability measure. Finally, from (4.58) it results by definition that:

$$\int_{A'(t)} S_i^*(t+1) dQ_{(t,A(t))} = S_i^*(t), \forall A'(t) \in \mathcal{P}(t), \quad (4.61)$$

and by construction:

$$\int_{A'(t)} S_i^*(t+1)dQ = S_i^*(t). \tag{4.62}$$

Since each element $A(t)$ of the σ-algebra $\mathcal{F}(t)$ represents a collection of events $A(t)$ of $\mathcal{P}(t)$ and since the discounted price at time t is $\mathcal{F}(t)$-measurable, it results that:

$$E_Q\left[S_i^*(t+1)\right] = S_i^*(t), \tag{4.63}$$

and hence Q is an equivalent martingale measure.

In the multiperiod setting a contingent claim is a $\mathcal{F}(t)$-measurable non-negative pay-off. Similarly to the single period model, the value at time T of an attainable contingent claim $g(T)$ equals the terminal value $V(T)$ of its generating trading strategy θ. From equations (4.53) and (4.56) it follows that for every $t < T$:

$$V^*(t) = V(t)/B(t) = E_Q[V(T)/B(T) \mid \mathcal{F}(t)], \tag{4.64}$$

and therefore the value at time t of a marketable contingent claim, $g(t)$, is given with:

$$g(t) = B(t)E_Q[g(T)/B(T) \mid \mathcal{F}(t)]. \tag{4.65}$$

By generalising the single period model it can be shown that the multiperiod model is complete if, and only if, there exists a unique equivalent measure Q.

Continuous-Time Model

A rigourous treatment of arbitrage in continuous time requires sophisticated mathematical tools and adds little to the economic insight. Continuous-time modelling offers however elegant pricing formulas for contingent claims. A detailed discussion of the continuous-time model can be found in Bingham and Kiesel (1998), Chap. 6. As for the discrete time model, the arbitrage price process is given by:

$$g(t) = B(t)E_Q[g(T)/B(T) \mid \mathcal{F}(t)]. \tag{4.66}$$

In continuous time the price processes are generally modelled as Itô processes (see Sect. 3.6):

$$dS(t) = a(w_t, t)dt + b(w_t, t)dW(t), \tag{4.67}$$

where $\{W(t)\}$ is a Wiener process. A central result, the *Girsanov's theorem*, provides conditions under which an equivalent measure exists. Let's consider the following process:

$$L(t) = \exp\left[-\int_0^t \gamma(u)dW(u) - \frac{1}{2}\int_0^t \gamma(u)^2 du\right], \qquad (4.68)$$

where $\gamma(t)$ with $0 \le t \le T$ is adapted and satisfies the following (Noikov's) condition:

$$E\left[\exp\left[\frac{1}{2}\int_0^T \gamma(u)^2 du\right]\right] < \infty. \qquad (4.69)$$

From the Girsanov's theorem it results that following process:

$$\hat{W}(t) = W(t) + \int_0^t \gamma(u)du, \qquad (4.70)$$

is a Wiener process under an equivalent measure Q with *Radon-Nikodým* derivative $dQ/dP = L(T)$ (see Sect. A.1). As it will be discussed in Sect. 4.4.2, the use of the Girsanov's theorem allows to change arbitrarily the drift of a given Itô process by redefining the process under an equivalent probability measure. Of particular interest is the redefinition of the discounted price process $\{S(t)\}$ under the risk-neutral probability measure Q.

4.2 Cash Instruments

4.2.1 Equities

An *equity* may be modelled as an infinite set of uncertain cash-flows (dividends). The forecast of cash-flows over long periods is, however, generally very imprecise. For this reason equities have been modelled as risk factors in Chap. 3. Given the corresponding risk factor $w_E(T)$, the valuation $g_E[w_E(T)]$ of each equity position θ_E is trivial:

$$g_E[w_E(T)] = \theta_E w_E(T), \qquad (4.71)$$

and linear in the risk factor.

A practical problem, however, is the risk assessment for large portfolios of stocks which typically involves hundreds of titles. Considering each stock as a risk factor leads to a very high dimensionality. A possibility to reduce the dimensionality of the model is considering a linear factor (or index) model, which suggests that returns (and therefore prices) tend to move together:

$$R_i = \alpha_i + \beta_i R_M + \varepsilon_i, \qquad (4.72)$$

where R_i is the return on the ith asset, R_m the rate of return on a factor (generally a market index), β_i represents the sensitivity of R_i to the factor,

α_i and ε_i are the components of R_i, which are independent of R_m. α_i is a constant and ε_i denotes a random error term whose expectation is zero by construction. Mathematically, (4.72) is obviously a linear regression where R_i is the dependent variable and R_m the explanatory variable and has the following properties (see for instance Jobson (1991), p.123):

- the distribution of the error term ε_i is independent of the distribution of the factor and hence $E[\varepsilon_i(R_m - E[R_m])] = 0$,
- the error terms are mutually independent and therefore $E[\varepsilon_i\varepsilon_j] = 0$ for $i, j \in I$ and $i \neq j$,
- the parameters α_i and β_i are constant.

Depending of the nature of the portfolio, the single index in (4.72) can be extended to a multi-index model in order to capture additional co-movements of the returns. A single index is probably appropriate for well-diversified passive index tracking portfolios. Multi-index models capture more adequately the investment style and therefore the risk of actively managed portfolios.

4.2.2 Fixed-Income Instruments

A *fixed-income instrument* (*bond*) consists of a set of certain cash-flows (if default risk is not considered) with different maturities. The value of a bond is therefore equal to the sum of the cash flows discounted by an appropriate rate. The valuation of fixed income instruments requires the modelling of the *term structure of interest rates*.

A bond which makes no periodic coupon payment and pays a certain amount B at maturity T is called a zero coupon bond (zero bond, discount bond). B is referred to as the face value or principal. The notation of Ingersoll (1987) will be used to describe the term structure of interest rates.

Let's define:

$P(t,T)$:   the present value at time t of one unit at time T ($t \leq T$), which is equal to the price of a zero bond with a nominal of one.

$Y(t,T)$:   the yield-to-maturity, i.e. the $T - t$ period interest rate prevailing at time t. The interest rate is stated on a period basis and compounded once in a given period of time.

r_t: the spot rate, the one period risk-free rate prevailing at time t for repayment at time $t + 1$.

$f(t,T)$:   the one period forward rate which is implied in the time t prices.

Following relations hold by definition:

$$f(t,T) = \frac{P(t,T)}{P(t,T+1)} - 1,$$

$$f(t,t) = Y(t,t+1) = r_t, \qquad (4.73)$$

$$P(t,T) = \frac{1}{(1+r_t)(1+f(t,t+1))\cdots(1+f(t,T-1))}.$$

In the context of certainty, i.e. for a given realisation of the risk factors describing the term structure, the no arbitrage condition requires that:

$$f(t,T) = r_T. \qquad (4.74)$$

If this condition is not met, an arbitrage profit through trading in T and $T+1$ can be realised. If $f(t,T) > (<)\, r_T$, an arbitrageur would sell short (buy) one bond maturing at time T and buy (sell) $1 + f(t,T)$ bonds at time $T+1$. The cost of the transaction is zero. At time T the investor would owe (own) one unit of currency and their long (short) position would be worth $\frac{1+f(t,T)}{1+r_T} > (<)\, 1$.

Thus the price of the zero bond with face value of one can be rewritten as:

$$P(t,T) = \frac{1}{(1+r_t)(1+r_{t+1})\cdots(1+r_{T-1})}. \qquad (4.75)$$

The dynamics of the interest rates are often modelled in continuous-time. The corresponding notation is:

$$P(t,T) = \exp(-Y(t,T)(T-t)),$$

$$f(t,t) = -\frac{1}{P(t,T)}\frac{\partial P(t,T)}{\partial T}, \qquad (4.76)$$

$$f(t,t) = r(t) = \lim_{T\to t} Y(t,T).$$

Under certainty, the continuous-time equivalent of (4.75) is:

$$P(t,T) = \exp\left[-\int_t^T r(s)ds\right]. \qquad (4.77)$$

Arbitrage arguments may be used to valuate zero bonds under uncertainty. Let's define the bank account process in continuous time with $B(0,T)$:

$$B(0,T) = \exp\left[\int_0^T r(s)ds\right]. \qquad (4.78)$$

Given the market is complete, it exists an equivalent martingale measure Q so that (see 4.66):

$$P(t,T) = B(0,t)E_Q\left[\frac{P(T,T)}{B(0,T)} \mid \mathcal{F}(t)\right],$$
$$= B(0,t)E_Q\left[\frac{P(T,T)}{B(0,t)B(t,T)} \mid \mathcal{F}(t)\right]. \tag{4.79}$$

Since at time t, $B(0,t)$ is deterministic, (4.79) may be rewritten (see Rebonato (1998), p.166):

$$P(t,T) = E_Q\left[\frac{P(T,T)}{B(t,T)} \mid \mathcal{F}(t)\right]. \tag{4.80}$$

As $P(T,T) = 1$ the value of $P(t,T)$ is therefore:

$$P(t,T) = E_Q\left[\exp\left[-\int_t^T r(s)ds\right] \mid \mathcal{F}(t)\right]. \tag{4.81}$$

Coupon bonds can be modelled as portfolios of zero bonds. A coupon bond $B(t,T)$ with face value B, known payments (C per period) and maturity T can be considered as a portfolio of zero bonds with face value C each maturing on one of the different coupon-payment dates plus a zero bond with maturity T and principal B:

$$B(t,T) = CP(t,t_1) + CP(t,t_2) \cdots (C+B)P(t,T). \tag{4.82}$$

4.3 Futures and Forwards

A *forward contract*, like a *futures contract* is an agreement between a buyer and a seller to trade an underlying asset at a future date for a specified (*delivery*) price. Unlike forward contracts, future contracts are generally standardised and traded on organised exchanges. Moreover, the institutional features of the futures markets are designed to reduce substantially the credit risk of both contracting parties. Usually an *initial margin* is required from both parties when the contract is initiated. The futures contracts are marked to market and are settled on a daily basis. The gain or loss is added to, respectively substracted from, the margin account of each party. If the margin account falls below a certain level, the *maintenance margin*, the corresponding party is required to restore the balance to the level of the initial margin.

As a consequence of the daily settlement, the price of a futures contract is not equal to the price of a forward contract with the same conditions. However, it can be shown that the prices of both contracts are equivalent if interest rates are constant and the term structure is flat (see Cox et al. (1981)). These conditions are not met in practice but the difference in price is usually small. It will therefore be assumed that the prices are equivalent and the issue of settlement will not be further analysed. A market without transaction costs, taxes and with the possibility of unlimited lending and borrowing at the constant risk-free rate is considered.

For the sake of generality, consider a somewhat unrealistic asset w with value $w(t)$ at time t which produces a known continuous dividend yield q and known cash income whose future value at T is I. It is assumed that the income from the w is reinvested in the asset so that a portfolio consisting of $e^{-q(T-t)}$ of the asset bought at time t has value $w(T)+I$ at time T. Table 4.1 compares this portfolio holding with a portfolio composed of a long forward contract f on w with maturity T and delivery price K and $(I+K)e^{-r(T-t)}$ of cash. It is assumed that the cash can be invested at a constant continuous risk-free rate r. Both strategies give the same pay-off at time T for all states of the world and therefore from the law of one price should have the same value at time t. It results that:

$$f(t) = w(t)e^{-q(T-t)} - (K+I)e^{-r(T-t)}. \tag{4.83}$$

The delivery price F for wich the price of a forward contract is zero at initiation is referred to as the *forward price*. From (4.83) the forward price is found to be:

$$F = w(t)e^{(r-q)(T-t)} - I. \tag{4.84}$$

Portfolio	Current value	Value at maturity
$e^{q(T-t)}$ of the asset	$w(t)e^{-q(T-t)}$	$w(T)+I$
One long forward and $(I+K)e^{-r(T-t)}$ of cash	$f(t)$ + $(K+I)e^{-r(T-t)}$	$w(T)-K$ + $K+I$

Table 4.1. Pricing of a futures contract

Note that under the assumptions made, both strategies produce the same pay-off at each time t with $t \in [0,T]$. By holding a positive position in one portfolio and a corresponding negative position in the other portfolio, the risk can be eliminated over the duration of the contract without the need of re-balancing the positions. Moreover, no assumption has been made about the price process of the risky asset w.

The forward price is a linear function of the underlying price. It can be modelled as a long position in the underlying and a short position in cash.

However, if the assumptions of a constant risk-free rate or known dividend yield are relaxed, an exact valuation requires the modelling of the interest rate or of the dividend process.

Given the general equation (4.83), the price of a forward contract on an asset that provides only a known cash income is given with:

$$f(t) = w(t) - (K + I)e^{-r(T-t)}. \tag{4.85}$$

Remark that I can be negative. For commodities for instance, the holding of the underlying asset generates storage costs.

The price of a forward contract on an asset which provides only a known dividend yield (as often assumed for market index) is:

$$f(t) = w(t)e^{-q(T-t)} - Ke^{-r(T-t)}. \tag{4.86}$$

This equation can be rewritten for forwards on currencies. The yield for the holder of the currency is assumed to be the risk-free rate of the foreign currency r_f. Therefore, the price of a forward contract becomes:

$$f(t) = w(t)e^{-r_f(T-t)} - Ke^{-r(T-t)}. \tag{4.87}$$

The pricing of bond futures where the physical delivery of the bond at maturity is possible faces a technical problem. The party who is short the futures has the option of delivering any bond within a given maturity range. Generally the price of the futures is derived from the price of the *cheapest to deliver bond*.

4.4 Options

This section focusses on the generic type of financial options, the European-type call (put) option, which is a contract with following conditions (see Wilmott et al. (1994), p.1):

- At a prescribed time in the future, known as the *expiry* or *expiration date*, the holder of the option may
- purchase (sell) a prescribed asset, known as the *underlying asset* or, briefly, the *underlying*, for a
- prescribed amount, known as the *exercise price* or *strike price*.

Contrary to futures, options give the owner the right not the obligation to purchase or sell the underlying asset. American-type options can be exercised at any time prior to expiry. Arbitrage-based arguments comparable to those used for the derivation of the future prices (without assumption on the distribution of the underlying assets) are extensively presented by Cox and Rubinstein (1985), Chap. 4 and by Spremann (1996), pp.628-638, for both call and put options. These arguments provide lower and upper bounds for the derivative prices. The first exact pricing formula is the well-known Black-Scholes formula, which is presented in the next section.

4.4.1 The Black-Scholes Analysis

The following assumptions are necessary to derive the Black-Scholes formula:

- The underlying price process $\{w(t)\}$ follows a geometric Brownian motion:

$$dw(t) = \mu w(t)dt + \sigma w(t)dW(t). \tag{4.88}$$

- The underlying asset pays no dividends during the life of the option.

- They are no riskless arbitrage opportunities.

- Trading of the underlying is continuous.

- Short selling is permitted and the assets are infinitely divisible.

- There are no transaction costs.

- The risk free rate is constant over the life of the option.

The striking idea of Black and Scholes (1973) used to derive their famous partial differential equation was to construct a portfolio containing a position in the option and in the underlying in order to remove the randomness. Consider an European option whose value $D(w, t)$ depends only on w and t. Using Itô's Lemma and dropping the arguments (see Hull (2000), p.246):

$$dD = \left(\mu w \frac{\partial D}{\partial w} + \frac{\partial D}{\partial t} + \frac{1}{2}\sigma^2 w^2 \frac{\partial^2 D}{\partial w^2} \right) dt + \sigma w \frac{\partial D}{\partial w} dW. \tag{4.89}$$

Let's construct a simple portfolio consisting of one option and a number $-\Delta$ of the underlying and whose value is Π. By definition:

$$\Pi = D - \Delta w, \tag{4.90}$$

and the change in value is therefore with Δ held fixed:

$$d\Pi = dD - \Delta dw, \tag{4.91}$$

which can be reexpressed as:

$$d\Pi = \left(\mu w \frac{\partial D}{\partial w} + \frac{\partial D}{\partial t} + \frac{1}{2}\sigma^2 w^2 \frac{\partial^2 D}{\partial w^2} - \mu \Delta w \right) dt + \sigma w \left(\frac{\partial D}{\partial w} - \Delta \right) dW. \tag{4.92}$$

as the Wiener processes underlying D and w are the same. By choosing

$$\Delta = \frac{\partial D}{\partial w}, \qquad (4.93)$$

the random component is eliminated and the increment in the portfolio value is deterministic:

$$d\Pi = \left(\frac{\partial D}{\partial t} + \frac{1}{2}\sigma^2 w^2 \frac{\partial^2 D}{\partial w^2} \right) dt. \qquad (4.94)$$

This means that the portfolio is perfectly hedged during dt and by using simple arbitrage arguments it follows that:

$$d\Pi = r\Pi dt, \qquad (4.95)$$

where r is the risk-free return. If the return of the portfolio were greater than $r\Pi dt$, arbitrageurs could make a guaranteed riskless profit by shorting risk-free securities to invest in the portfolio. The converse argument is naturally true if the return of the portfolio is less than $r\Pi dt$. By substitution it follows that:

$$\frac{\partial D}{\partial t} + \frac{1}{2}\sigma^2 w^2 \frac{\partial^2 D}{\partial w^2} + rw\frac{\partial D}{\partial w} - rD = 0, \qquad (4.96)$$

which is the Black-Scholes partial differential equation. This equation plays a central role in the pricing of derivatives. Any derivative instrument whose value depends only on the current value of w and t must satisfy this equation or a variant incorporating dividends or time dependent parameters (see Wilmott et al. (1994)).

Mathematically, the Black-Scholes equation is a parabolic equation with many solutions. To obtain a unique value boundary conditions must be defined. Two conditions must be posed in w (due to the second derivative associated with it) and one in t.

The final condition (see Wilmott et al. (1994), pp.46-48) is the value of the derivative instrument at time $t = T$, which is known with certainty. For a call whose value is denoted by $C(w, t)$ the pay-off is:

$$C(w, T) = \max(w(T) - K, 0), \qquad (4.97)$$

where K is the strike price. For a put with value $P(w, t)$ the final condition yields:

$$P(w, T) = \max(K - w(T), 0). \qquad (4.98)$$

The conditions in w, the spatial or asset-price boundaries, are the prices of the derivative as $w = 0$ and $w \to \infty$. As w follows a geometric Brownian motion, it has an absorbing barrier at 0. Thus, if w hits 0, it will remain at 0. Therefore, the call option is and will remain worthless until expiry. The put will be exercised and its final pay-off is known with certainty and is equal to K. Its value at time t is the present value of K. Hence, for r constant:

$$C(0,t) = 0,$$
$$P(0,t) = e^{-r(T-t)}K. \tag{4.99}$$

As the call is a right to buy the underlying it can never be worth more than the underlying. As $w \to \infty$ the call reaches its upper boundary. As the underlying price increases without bound, it becomes very unlikely that the put will be exercised and therefore:

$$C(w,t) \approx w \text{ as } w \to \infty,$$
$$P(w,t) \to 0 \text{ as } w \to \infty. \tag{4.100}$$

Once the boundary conditions are posed, the solution of the Black-Scholes equation is unique and yields following well-known formula for the European call:

$$C(w,t) = w(t)\Phi(d_1) - Ke^{-r(T-t)}\Phi(d_2), \tag{4.101}$$

respectively for the corresponding put:

$$P(w,t) = Ke^{-r(T-t)}\Phi(-d_2) - w(t)\Phi(-d_1), \tag{4.102}$$

where:

$$d_1 = \frac{\log(w(t)/K) + (r + \sigma^2/2)(T-t)}{\sigma\sqrt{T-t}},$$
$$d_2 : d_1 - \sigma\sqrt{T-t},$$

and Φ denotes the cumulative probability distribution function of the standard normal distribution.

4.4.2 Risk-Neutral Valuation

The pricing of standard European put and call options by risk-neutral valuation is a direct application of the results discussed in Sect. 4.1.1. Since the price of the equity $\{w(t)\}$ in the Black-Scholes framework is modelled by following stochastic differential equation:

$$dw(t) = \mu w(t)dt + \sigma w(t)dW(t), \tag{4.103}$$

and the deterministic bank account process $\{B(t)\}$, with initial value $B(0) = 1$, is represented by the following dynamics:

$$dB(t) = rB(t)dt, \tag{4.104}$$

it results directly from Itô's Lemma that the dynamics of the discounted price process $\{w^*(t)\}$, $w^*(t) = w(t)/B(t)$, are given with:

$$dw^*(t) = (\mu - r)w(t)^* dt + \sigma w^*(t)dW(t). \tag{4.105}$$

From the Girsanov's theorem it follows that if Q and P are equivalent probability measures, there exists a Wiener process $\{\hat{W}(t)\}$ under Q so that (see (4.70)):

$$dW(t) = d\hat{W}(t) - \gamma(t)dt, \tag{4.106}$$

and (4.105) may be rewritten:

$$dw^*(t) = (\mu - r - \sigma\gamma(t))w(t)^* dt + \sigma w^*(t)d\hat{W}(t). \tag{4.107}$$

If Q is the martingale measure the discounted price process has a drift of zero and therefore:

$$\mu - r - \sigma\gamma(t) = 0, \tag{4.108}$$

which implies:

$$\gamma(t) = \frac{\mu - r}{\sigma}. \tag{4.109}$$

By applying Itô's Lemma to (4.107), it is easy to verify that the price process $\{w(t)\}$ dynamics are given by:

$$dw(t) = rw(t)dt + \sigma w(t)d\hat{W}(t), \tag{4.110}$$

which represents a geometric Brownian motion with drift equal to the risk-free rate. From equations (4.107) and (4.108) it results that the value at time t of the discounted price process yields:

$$w^*(t) = w^*(0)\exp\left[-\frac{1}{2}\sigma^2 t + \sigma\hat{W}(t)\right]. \tag{4.111}$$

Note that the bank account process is deterministic and therefore $B(T) = \exp(rT)$ holds also under Q. From (4.65) the actual price $C(0)$ of a European call may be written:

$$C(0) = B(0)E_Q[C(T)/B(T) \mid \mathcal{F}(0)],$$
$$= E_Q\left[\max\left(w^*(0)\exp\left[-\frac{1}{2}\sigma^2 T + \sigma\hat{W}(T)\right] - \exp[-rT]K, 0\right)\right]. \tag{4.112}$$

Using the fact that $\{\hat{W}(t)\}$ is a Wiener process under Q and therefore $\hat{W}(T) \sim \mathcal{N}(0, T)$ it can be verified that (4.112) is equivalent to the Black-Scholes formula (4.101). A derivation is given in Sundaram (1997).

Note that the Black-Scholes approach requires the computation of a specific replication strategy for every derivative. The risk-neutral valuation on the other hand allows the pricing of every contingent claim by using the same scheme, namely by discounting the expected value of its expected pay-off at maturity under Q.

4.4.3 Numerical Approaches

The Black-Scholes formula was a major breakthrough and led to a rapid growth of the option market. The model has been generalised and considerably extended. The generalisations have relaxed the assumptions originally made and the extensions have developed the model for American options, for alternative one- and multidimensional price processes, and for options on alternative pay-off patterns which are generally referred to as exotic options. For example the pay-off of *barrier options* depends on whether the underlying's price reaches a certain level within a given period. The option can either come into existence or become worthless if the level is reached. The pay-off of a *lookback option* depends on the maximum or minimum underlying's price during the life of the option. *Asian options* are derivatives whose pay-off depends on some form of the average price of the underlying asset. This review is by no means exhaustive. More important than a listing of all available option variants is the fact that for most of them no closed-form pricing formula is available. An approximation of their values is obtained by means of numerical methods. The great advantage of the numerical procedures are their flexibility and their relative simplicity. Numerical methods are, however, computationally demanding as they have to discretise the probability space of the underlying asset(s). They are widespread for pricing purposes but their implementation in the context of risk management where large portfolios of derivatives have to be monitored simultaneously remain challenging. Three classes of numerical procedures are generally used to value options. The *finite difference methods* are a means to obtain numerical solutions for partial differential equations. The *lattice models* and *the Monte Carlo simulation* are applications of the risk-neutral valuation.

Finite Difference Methods

The idea underlying the finite difference methods is to replace the partial derivatives by finite differences. This approach was introduced by Schwartz (1977) and Brennan and Schwartz (1978) in the context of derivatives pricing. For the Black-Scholes equation, the two dimensions, time and underlying price, have to be discretised. The time axis $(t = [O, T])$ is divided into $N + 1$ equally spaced nodes a distance $\Delta t = T/N$ apart. (see Wilmott et al. (1994)) Similarly, on the space axis $M + 1$ equally spaced underlying prices w between 0 and w_{max} are considered with $\Delta w = w_{max}/M$. Let $D_{n,m}$ denote the value of the derivative instrument at the point $(n\Delta t, m\Delta w)$. w_{max} is a large value of w for which $D(t, w_{max}) \approx D(t, \infty)$. the support of the discretisation consists therefore on the $(N + 1)(M + 1)$ points of the form $(n\Delta t, m\Delta w)$ with $n = 0, \cdots, N$ and $m = 0, \cdots, M$. For an interior point, i.e. a point $(n\Delta t, m\Delta w)$ with $n < N$ and $m < M$ the partial derivatives of $\frac{\partial D}{\partial w}$ for instance can be approximated with the *forward difference*:

$$\frac{\partial D}{\partial w} \approx \frac{D_{n,m+1} - D_{n,m}}{\Delta w}, \tag{4.113}$$

the *backward difference*:

$$\frac{\partial D}{\partial w} \approx \frac{D_{n,m} - D_{n,m-1}}{\Delta w}, \tag{4.114}$$

or alternatively the *central difference*:

$$\frac{\partial D}{\partial w} \approx \frac{D_{n,m+1} - D_{n,m-1}}{2\Delta w}. \tag{4.115}$$

The central difference approximations exhibit quadratic convergence whereas the forward and backward approximations exhibit linear convergence. The central differences are therefore more accurate for small Δw. The finite-difference approximations of $\frac{\partial D}{\partial t}$ and $\frac{\partial^2 D}{\partial w^2}$ are defined the same way (There are $3^3 = 27$ possible approximations for the second partial derivative $\frac{\partial^2 D}{\partial w^2}$). Central differences are generally not used for $\frac{\partial D}{\partial t}$ because they lead to bad numerical schemes (see Wilmott et al. (1994), p.270).

Using the forward difference approximation for $\frac{\partial D}{\partial t}$ and the central difference approximation for $\frac{\partial D}{\partial w}$ and $\frac{\partial^2 D}{\partial w^2}$, the Black-Scholes equation may be rewritten:

$$\frac{D_{n+1,m} - D_{n,m}}{\Delta t} + rm\Delta w \frac{D_{n,m+1} - D_{n,m-1}}{2\Delta w}$$
$$+ \frac{1}{2}\sigma^2 m^2 \Delta w^2 \frac{D_{n,m+1} - 2D_{n,m} + D_{n,m-1}}{\Delta w^2} = rD_{n,m} \tag{4.116}$$

for $n = 0, \cdots, N - 1$ and $m = 0, \cdots, M - 1$

By rearranging the terms it results that (see Hull (2000), p.419):

$$a_m D_{n,m-1} + b_m D_{n,m} + c_m D_{n,m+1} = D_{n+1,m}. \tag{4.117}$$

where:

$$a_m = \frac{1}{2}rm\Delta t + \frac{1}{2}\sigma^2 m^2 \Delta t.$$

$$b_m = 1 + \sigma^2 m^2 \Delta t + r\Delta t. \qquad (4.118)$$

$$c_m = -\frac{1}{2}rm\Delta t + \frac{1}{2}\sigma^2 m^2 \Delta t.$$

The equation above and the boundary conditions for $t = T$, $w = 0$ and $w = w_{max}$ allow to solve the $M - 1$ equations for the points $(T - \Delta t, m\Delta w)$, $m = 1, \cdots M - 1$. The next step consists in solving the $M - 1$ equations for the points $(T - 2\Delta t, m\Delta w)$, $m = 1, \cdots M - 1$ and so on until $t = 0$ is reached. Hence, N systems of $M - 1$ simultaneous equations have to be solved. This procedure is referred to as the *implicit difference method*. A simplified approach is the *explicit finite difference* method where the approximations of $\frac{\partial D}{\partial w}$ and $\frac{\partial^2 D}{\partial w^2}$ for the points (n, m) and $(n + 1, m)$ are assumed to be equal. The equation reduces to:

$$D_{n,m} = a'_m D_{n+1,m-1} + b'_m D_{n+1,m} + c'_m D_{n+1,m+1}, \qquad (4.119)$$

where:

$$a'_m = \frac{1}{1 + r\Delta t}\left(-\frac{1}{2}rm\Delta t + \frac{1}{2}\sigma^2 m^2 \Delta t\right).$$

$$b'_m = \frac{1}{1 + r\Delta t}\left(1 - \sigma^2 m^2 \Delta t\right). \qquad (4.120)$$

$$c'_m = \frac{1}{1 + r\Delta t}\left(\frac{1}{2}rm\Delta t + \frac{1}{2}\sigma^2 m^2 \Delta t\right).$$

The explicit finite difference method does not require the solution of systems of equation. However, the method has poorer convergence properties than the implicit finite difference method (see Wilmott et al. (1994), p.305).

The finite difference methods are particularly useful for derivatives where some decisions (like early exercise) have to be made prior maturity and are therefore extensively used for American options.

The Lattice Models

The lattice models begin with a discretisation of the time axis. At any time step the underlying price can move from its level at the start of the time step to a finite number of new levels at the end of the time step. The binomial model allows for two new levels whereas the trinomial approach models three possible future movements.

The Binomial Model The binomial model was suggested by Sharpe (1978) and derived by Cox et al. (1979). The binomial model is a direct application of the risk-neutral valuation model in a discrete-time, multiperiod setting. In this approach, the time axis is discretised by dividing the life of the option into N time intervals Δt. The underlying price process $\{w(t)\}$ is supposed to change at the discrete times $\Delta t, 2\Delta t \cdots N\Delta t$. The initial price of the underlying is equal to $w(0)$. At date Δt the price can amount $dw(0)$, respectively $uw(0)$, with probability p, respectively $1 - p$. From $dw(0)$, respectively $uw(0)$, the underling price can transit to price $d^2 w(0)$, respectively $duw(0)$, with conditional probability p or to price $duw(0)$, respectively $u^2 w(0)$, with conditional probability $1 - p$. At date $n\Delta t$ the price of the underlying is for arbitrary integers $n_1, n_2 \in [0, n]$ given that $n_1 + n_2 = n$ equal to:

$$d^{n_1} u^{n_2} w(0), \tag{4.121}$$

with corresponding probability:

$$p^{n_1}(1 - p)^{n_2}. \tag{4.122}$$

Clearly , the conditional mean and variance of $w((n + 1)\Delta t)$, given $w(n\Delta t)$, are:

$$E\big[w((n + 1)\Delta t) \mid w(n\Delta t)\big] = puw(n\Delta t) + (1 - p)dw(n\Delta t),$$
$$\mathrm{var}\big[w((n + 1)\Delta t) \mid w(n\Delta t)\big] = pu^2 w^2(n\Delta t) + (1 - p)d^2 w^2(n\Delta t) \tag{4.123}$$
$$- [pu + (1 - p)d]^2 w^2(n\Delta t).$$

From equations (4.121) and (4.122) it results that the underlying price process is assumed to follow multiplicative binomial process which converges to the geometric Brownian motion as $N \to \infty$. As shown in Sect. 4.4.2, under risk neutrality the corresponding process is geometric Brownian motion with drift equal to the risk-free rate r (see 4.110):

$$\frac{dw(t)}{w(t)} = rdt + \sigma d\hat{W}(t), \tag{4.124}$$

with following conditional mean and variance:

$$E\big[w((n + 1)\Delta t) \mid w(n\Delta t)\big] = w(n\Delta t)e^{r\Delta t},$$
$$\mathrm{var}\big[w((n + 1)\Delta t) \mid w(n\Delta t)\big] = w^2(n\Delta t)e^{2r\Delta t}(e^{\sigma^2 \Delta t} - 1), \tag{4.125}$$

The parameters u, d and p of the binomial method are generally calibrated so that the expected values and variances of the underlying price are equal after each time step Δt to the corresponding statistics of the risk-neutral geometric Brownian motion:

$$pu + (1 - p)d = e^{r\Delta t},$$
$$pu^2 + (1 - p)d^2 - e^{2r\Delta t} = e^{2r\Delta t}(e^{\sigma^2 \Delta t} - 1). \tag{4.126}$$

Equations (4.126) provide two conditions for the three unknown parameters u, d and p. A third condition may be chosen arbitrary. The original choice of Cox et al. (1979) is:

$$u = \frac{1}{d}, \tag{4.127}$$

so that after two time steps the underlying price can return to its initial value. Solving with the conditions given by equations (4.126) and (4.127) the values for u, d and p are found to be (see Hull and White (1988)):

$$p = \frac{e^{r\Delta t} - d}{u - d},$$

$$u = \frac{e^{2r\Delta t}e^{\sigma^2 \Delta t} + 1 + \sqrt{\left(e^{2r\Delta t}e^{\sigma^2 \Delta t} + 1\right)^2 - 4e^{2r\Delta t}}}{2e^{r\Delta t}}, \tag{4.128}$$

$$d = \frac{1}{u}.$$

The rather cumbersome formulas for u and d are in practice replaced by the simpler expressions (see Hull (2000), p.390):

$$u = e^{\sigma\sqrt{\Delta t}},$$
$$d = e^{-\sigma\sqrt{\Delta t}}, \tag{4.129}$$

which are obtained by expanding the u of (4.128) in Taylor series in powers of $\sqrt{\Delta t}$ and by neglecting terms of order Δt^2 and higher.

Note that for large time steps $r\Delta t > \sigma$ and therefore $p > 1$, respectively $1 - p < 0$. Hence, the model of Cox et al. (1979) allows for negative probabilities. To avoid this drawback, condition (4.127) can be replaced by (see for instance Jarrow and Rudd (1983)):

$$p = 0.5. \tag{4.130}$$

An approximative solution of (4.126) and (4.130) is obtained by neglecting terms of order Δt^2 and higher (see Hull (2000), p.403:)

$$u = e^{(r - \frac{\sigma^2}{2})\Delta t + \sigma\sqrt{\Delta t}},$$
$$d = e^{(r - \frac{\sigma^2}{2})\Delta t - \sigma\sqrt{\Delta t}}. \tag{4.131}$$

Once the discretisation of the underlying process is computed, possible values for the underlying price at maturity are obtained. Derivatives whose price at expiry depends only on the value of the underlying at T can be valuated for each realisation of w. For instance, the value of an European call is given with:

$$C^n(T) = \max[w^n(T) - K, 0], \tag{4.132}$$

where $C^n(T)$ is the value of the call given $w^n(T)$, the nth possible value of the underlying at expiration. As risk-neutrality is assumed, the value of the call at time $n\Delta t$ is equal to the expected value at time $(n+1)\Delta t$ discounted with the risk-free rate:

$$e^{r\Delta t}C^n_{n\Delta t} = p\,C^{n+1}_{(n+1)\Delta t} + (1-p)\,C^n_{(n+1)\Delta t}. \qquad (4.133)$$

Note that as μ and σ are constant, the tree recombines, i.e. an up movement followed by a down movement produces the same value as a down movement followed by an up movement. Therefore the number of lattice points is equal to $\sum_{n=1}^{N} n$. However, for alternative price processes, the tree will not necessarily recombine. In this case the number of nodes is equal to 2^N and the procedure is slow. In case the tree recombines, one node can be reached by several paths and so the history of a particular path is lost. Therefore the method is not optimal for path dependent options. Like the finite differences methods, the binomial tree procedure is particularly useful for the pricing of American options and relatively inefficient for the pricing of derivatives with several underlyings.

The Trinomial Model The trinomial model is an extension of the binomial model where the underlying price $w(n\Delta t)$, $(n = 0, 1, \cdots, N-1)$, can amount after a time step Δt three different values: $dw(n\Delta t)$, $mw(n\Delta t)$ and $uw(n\Delta t)$. The corresponding conditional probabilities are noted with p_d, p_m and p_u. Clearly, the following condition has to be satisfied:

$$p_d + p_m + p_u = 1, \qquad (4.134)$$

and according to (4.126):

$$\begin{aligned}
p_u u + p_m m + p_d d &= e^{r\Delta t}, \\
p_u u^2 + p_m m^2 + p_d d^2 - e^{2r\Delta t} &= e^{2r\Delta t}(e^{\sigma^2 \Delta t} - 1).
\end{aligned} \qquad (4.135)$$

Equations (4.134) and (4.135) impose three conditions on the six parameters d, m, u, p_d, p_m and p_u. Following further conditions are used frequently:

$$u = \frac{1}{d}, \qquad (4.136)$$

and:

$$m = 1. \qquad (4.137)$$

With condition (4.137) the trinomial model assigns a probability to the event "the underlying value remains fixed".

Parameter values for d, m, u, p_d, p_m and p_u which satisfy equations (4.134) to (4.137) can be found in Hull (2000), p.405:

$$u = e^{\sigma\sqrt{3\Delta t}},$$

$$p_d = -\sqrt{\frac{\Delta t}{12\sigma^2}}\left(r - \frac{\sigma^2}{2}\right) + \frac{1}{6},$$

$$p_m = \frac{2}{3}, \tag{4.138}$$

$$p_u = \sqrt{\frac{\Delta t}{12\sigma^2}}\left(r - \frac{\sigma^2}{2}\right) + \frac{1}{6}.$$

Another interessant derivation (see Wilmott et al. (1994), pp.406-407) can be found by considering the possible values of the underlying after two time steps in the binomial model:

$$
\begin{aligned}
w((n+1)\Delta t) &= u^2 w(n\Delta t) \quad &&\text{with probability } p^2, \\
w((n+1)\Delta t) &= w(n\Delta t) \quad &&\text{with probability } 2p(1-p), \\
w((n+1)\Delta t) &= d^2 w(n\Delta t) \quad &&\text{with probability } (1-p)^2,
\end{aligned} \tag{4.139}
$$

for $n = 0, 1, \cdots, N-1$. Using conditions (4.136) and (4.137) and replacing Δt by $\Delta t/2$ in (4.128) and (4.129) the parameters are given with:

$$
\begin{aligned}
u = e^{2\sigma\sqrt{\Delta t/2}}, \; m = 1, \; d = e^{-2\sigma\sqrt{\Delta t/2}}, \\
p_u = p^2, \; p_m = 2p(1-p), \; p_d = (1-p)^2,
\end{aligned} \tag{4.140}
$$

with:

$$p = \frac{e^{r\Delta t/2} - d}{u - d}. \tag{4.141}$$

Hence, a single step of the trinomial model is equivalent to two steps of the binomial model.

At every node the valuation of a derivative, whose possible values $D_{N\Delta t}$ at time T are known, is given with:

$$e^{r\Delta t} D_{n\Delta t} = p_u D^u_{(n+1)\Delta t} + p_m D^m_{(n+1)\Delta t} 1 + p_d D^d_{(n+1)\Delta t}, \tag{4.142}$$

where D^u, D^m, and D^d are the prices of the derivative instrument after the corresponding movements $uw(n\Delta t)$, $mw(n\Delta t)$ and $dw(n\Delta t)$ of the underlying and for $n = 0, 1, \cdots, T-1$.

It is worth noting that the trinomial model is closely related to the implicit finite difference approach. Equations (4.119) may be rewritten as follows:

$$(1 + r\Delta t) D_{n,m} = a^*_m D_{n+1,m-1} + b^*_m D_{n+1,m} + c^*_m D_{n+1,m+1}, \tag{4.143}$$

$$a_m^* = -\frac{1}{2}rm\Delta t + \frac{1}{2}\sigma^2 m^2 \Delta t.$$
$$b_m^* = 1 - \sigma^2 m^2 \Delta t. \qquad\qquad (4.144)$$
$$c_m^* = \frac{1}{2}rm\Delta t + \frac{1}{2}\sigma^2 m^2 \Delta t.$$

It is easy to verify that a_m^*, b_m^* and c_m^* sum to unity and can be interpreted as probabilities. The value of the derivative instrument at time $n\Delta t$ is therefore equal to the expected value at time $(n + 1)\Delta t$ discounted with the risk-free rate. As for the trinomial model the expected value at time $(n + 1)\Delta t$ of the derivative instrument is modelled as the average of three possible values weighted by their corresponding probabilities.

Monte Carlo Simulation

The Monte Carlo simulation methodology is discussed in greater details in the next chapter. For the pricing of derivatives the approach consists of the following steps (see Boyle et al. (1997)):

- simulate sample paths of the underlying variables over the relevant time horizon and according to the risk-neutral measure,
- evaluate the discounted cash flows of the security on each sample path,
- average the discounted cash flows over the sample paths.

Compared to the other numerical procedures, the Monte Carlo simulation is attractive for the pricing of derivatives on several underlyings. As discussed in the next chapter the error of the Monte Carlo estimate is independent of the dimension of the underlying space. Using the Monte Carlo simulation for the valuation of American options or more generally of options whose cash flows is influenced by decisions to be made prior maturity is not straightforward. A review of the recent developments can be found in Boyle et al. (1997).

4.5 Approximation of the Value Function

As discussed above, the computation of price functions can be rather time-demanding when numerical procedures have to be used. A standard procedure to overcome this difficulty consists in substituting the risk profile with its pth order Taylor approximation about a given point, generally about the origin $w(0)$, which will be called the supporting point of the approximation.

It is well known, that if $g : R^M \to R$ is $C^{p+1}(G)$, i.e. continuous and $p + 1$-times differentiable on a ball G about $x(0) \in \mathbb{R}^M$ $(x(0) = x_1(0), x_2(0), \cdots, x_M(0))$, there exists a C^{p+1} function $R_p[.; x(0)]$ so that for all points $x(0) + d \in G$ (see for instance Norton and Blume (1994), p.835):

$$g[x(0) + d] = g[x(0)] + \sum_{j_1=1}^{M} \frac{\partial g[x(0)]}{\partial x_{j_1}} d_{j_1} + \frac{1}{2!} \sum_{j_1=1}^{M} \sum_{j_2=1}^{M} \frac{\partial^2 g[x(0)]}{\partial x_{j_1} \partial x_{j_2}} d_{j_1} d_{j_2} + \cdots$$

$$\cdots + \sum_{j_1=1}^{M} \sum_{j_2=1}^{M} \cdots \sum_{j_p=1}^{M} \frac{\partial^p g[x(0)]}{\partial x_{j_1} \partial x_{j_2} \cdots \partial x_{j_p}} d_{j_1} d_{j_2} \cdots d_{j_p} + R_p[d; x(0)],$$

$$(4.145)$$

or in short from notation:

$$g[x(0) + d] = g[x(0)] + \sum_{i=1}^{p} \frac{1}{i!} \left[\sum_{j_1=1}^{M} \cdots \sum_{j_i=1}^{M} \frac{\partial^i g[x(0)]}{\partial x_{j_1} \cdots \partial x_{j_i}} \prod_{k=1}^{i} d_{j_k} \right] + R_p[d; x(0)],$$

$$(4.146)$$

where (see for instance Heuser (1993), p.282):

$$R_p[d; x(0)] = \frac{1}{p+1!} \sum_{j_1=1}^{M} \cdots \sum_{j_{p+1}=1}^{M} \frac{\partial^i g[x(0) + \vartheta d]}{\partial x_{j_1} \cdots \partial x_{j_i}} \prod_{k=1}^{p+1} d_{j_k}, \quad \vartheta \in [0, 1]. \quad (4.147)$$

Equation (4.145) is referred to as a p-order Taylor (series) expansion of g at $x(0)$. Without the remainder term $R_p[d; x(0)]$, the right-hand side of (4.145) is called a pth order Taylor (series) approximation of g about the supporting point $x(0)$. Note that the remainder $R_p[d; x(0)]$ satisfies:

$$\frac{R_p[d; x(0)]}{\|d\|^p} \to 0 \text{ as } d \to 0, \quad (4.148)$$

where $\|d\|$ denotes the length of d.

Clearly, for a given realisation $w(T) \in \mathbb{R}^M$ of the risk factors at time T, the pth order Taylor approximation of the risk profile about the origin $w(0)$ is given by (with $d = w(T) - w(0)$):

$$g[w(T)] \approx g[w(0)] + \sum_{i=1}^{p} \frac{1}{i!} \left[\sum_{j_1=1}^{M} \cdots \sum_{j_p=1}^{M} \frac{\partial^i g[w(0)]}{\partial w_{j_1} \cdots \partial w_{j_i}} d_{j_1} \cdots d_{j_i} \right]. \quad (4.149)$$

Since p, the order of the approximation, is finite, a Taylor approximation of the risk profile leads to a truncation error whose size is given by the remainder $R_p[d; w(0)]$. When the derivatives of order higher than p can not be neglected, the approximation performs well only locally, i.e. in the neighbourhood of the supporting point. This fact is frequently overlooked in the practice of risk management, where a single Taylor polynomial about $w(0)$ is usually used as a global approximation of the whole risk profile.

The numerical effort associated with a Taylor approximation of order p is directly related to the number of partial derivatives which have to be evaluated. From (4.149) it is easy to verify that a pth order Taylor approximation involves the computation of $\sum_{i=1}^{p} \frac{1}{i!} \prod_{k=1}^{i}(M + k - 1)$ partial derivatives. It is current practice in risk management to consider first or second order Taylor approximations. In the sequel, first order approximations will be called linear approximations while second order will be referred to as quadratic approximations. In case the risk profile is not sufficiently differentiable, partial derivatives can be obtained numerically: for a discussion see Press et al. (1992), pp.186-189 for instance.

4.5.1 Global Taylor Approximation for Option Pricing

JP Morgan (1996) analyses the accuracy of the quadratic approximation for the pricing of call and put options with the Black-Scholes formula as benchmark. The results show that the relative error depends on the relation of the spot and strike price and on the time to maturity. The error increases when the option approaches expiration at-the-money. An obvious explanation is offered by the non-differentiability of the risk profile at the strike price when the option expires. Another study on the the adequacy of Taylor polynomials for the approximation of option prices can be found in Estrella (1995), on which the following is based.

Let's consider a one-dimensional risk profile consisting of a single call option whose exact value is given by the Black-Scholes formula (see (4.101)). It is further assumed, without loss of generality, that the strike price and the time to maturity equal one, the risk-free interest rate is zero and the current time $t = 0$.

The price of the call (see (4.101) may therefore be rewritten as:

$$g[w(T)] = w(T)\Phi\left(\frac{\log(w(T)) + \frac{1}{2}\sigma^2}{\sigma}\right) - \Phi\left(\frac{\log(w(T)) - \frac{1}{2}\sigma^2}{\sigma}\right), \quad (4.150)$$

and the complete Taylor expansion for $g[w(T)]$ is given with the following power series:

$$g[w(T)] = \sum_{p=0}^{\infty} \frac{g^{(p)}w(0)}{p!} (w(T) - w(0))^p, \quad (4.151)$$

where $g^{(p)}$ denotes the pth order derivative of g.

It seems plausible at first glance to obtain an approximation with any desired degree of accuracy by choosing p sufficiently large. This is only true if the series converges. Note that a power series is differentiable at every point of its convergence interval and the convergence properties of any of its derivatives are the same as those of the series itself (see for instance Gélinas (1984), pp.42-43). As a result, the convergence of the right-hand side of (4.151) may be analysed by the means of its second derivative:

$$g^{(2)}[w(T)] = \frac{1}{\sigma w(T)} \Phi' \left(\frac{\log(w(T)) + \frac{1}{2}\sigma^2}{\sigma} \right) = \frac{e^{\frac{-1}{2\sigma^2} \left(\log(w(T)) + \frac{1}{2}\sigma^2 \right)^2}}{\sqrt{2\pi}\sigma w(T)}.$$

(4.152)

The expression is built from the exponential, the logarithmic, the identity, and the square functions. The exponential series, $w(T)^2$, and $w(T)$ are everywhere convergent. As a result, the Taylor approximation of $g[w(T)]$ converges over the range of convergence of $\log(w(T))$ (for a derivation, see Estrella (1995), appendix 1).

It is well known that (see for instance Gélinas (1984), p.51):

$$\log(1 + x) = \sum_{p=1}^{\infty} (-1)^{p-1} \frac{x^p}{p} \qquad (4.153)$$

converges for $\mid x \mid < 1$ and diverges for $\mid x \mid > 1$. Therefore, $\log(w(T))$ converges over the range $0 < w(T) < 2$.

Figures 4.1 and 4.2 depict the Taylor approximations up to the 20th order of the risk profile for an holding period of one year ($T = 1$) and an important but not unrealistic positive shock of the market price of three standard deviations, i.e. $w(T)$ is given with $e^{3\sigma w(0)}$. For $w(0) = 1$ and $\sigma = 25\%$, $w(T)$ will amount approximatively 2.117 and the Taylor series does not converge to the exact price. As illustrated in Fig. 4.1, the error of the approximation tends to increase with the number of terms considered.

For $w(0) = 1$ and $\sigma = 5\%$, $w(T)$ is about 1.162 and the series converges. However, as shown in Fig. 4.2 only high order approximations achieve a good level of accuracy. Here, a minimum order of 15 is required to obtain an error no greater than 5%. The important implication is that for large market moves global Taylor approximations may perform very poorly. This is a particular concern for long holding horizons where large moves are more likely, as well as for the estimation of risk measures focussing on tail events like VaR or TailVaR.

4.5.2 Piecewise Taylor Approximations

The lack of convergence, respectively the need for high order terms in the event of large market moves, challenges the adequacy of global Taylor approx-

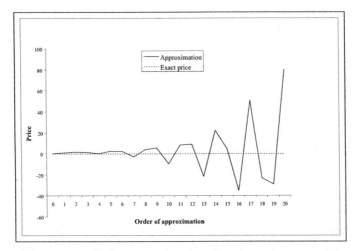

Figure 4.1. Taylor approximation of option price, $\sigma = 25\%$

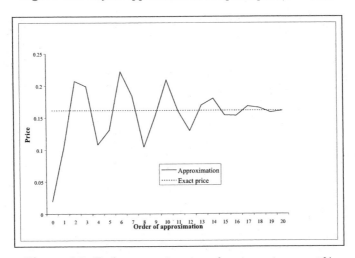

Figure 4.2. Taylor approximation of option price, $\sigma = 5\%$

imations. Moreover, terms higher than the second are generally not available in the practice of risk management. One possible way is to replace the global approach by a piecewise one, i.e. to use several supporting points to compute the Taylor approximation.

Let's consider the above example with $\sigma = 25\%$. Table 4.2 compares the results of the full valuation with three quadratic approximations for market moves in the range $[-3\sigma, +3\sigma]$. The first approximation is global around 0. The second one is piecewise with supporting points -1.5σ and $+1.5\sigma$. Finally, the third approximation uses the three supporting points $-2\sigma, 0, +2\sigma$.

Market Shock	Full valuation	Quadratic approx. global		Quadratic approx. piecewise with 2 s.p.	
	$g(w)$	$\hat{g}(w)$	% error	$\hat{g}(w)$	% error
-3σ	0.000065	0.029816	45641.93%	0.008701	13248.57%
-2.5σ	0.000364	0.014980	4013.62%	0.003609	891.14%
-2σ	0.001643	0.005737	249.17%	0.002152	31.00%
-1.5σ	0.006040	0.004984	-17.48%	0.006040	0.00%
-1σ	0.018290	0.016611	-9.18%	0.017556	-4.01%
-0.5σ	0.046269	0.045811	-0.99%	0.039737	-14.12%
0σ	0.099477	0.099477	0.00%	0.076595	-23.00%
0.5σ	0.185578	0.186709	0.61%	0.173868	-6.31%
1σ	0.307511	0.319481	3.89%	0.305998	-0.49%
1.5σ	0.463779	0.513493	10.72%	0.463779	0.00%
2σ	0.651430	0.789272	21.16%	0.652916	0.23%
2.5σ	0.868926	1.173590	35.06%	0.880521	1.33%
3σ	1.117138	1.701299	52.29%	1.155490	3.43%
Market Shock	Full valuation	Quadratic approx. global		Quadratic approx. piecewise with 3 s.p.	
	$g(w)$	$\hat{g}(w)$	% error	$\hat{g}(w)$	% error
-3σ	0.000065	0.029816	45641.93%	0.001648	2427.76%
-2.5σ	0.000364	0.014980	4013.62%	0.000629	72.6%
-2σ	0.001643	0.005737	249.17%	0.001643	0.00%
-1.5σ	0.006040	0.004984	-17.48%	0.005577	-7.66%
-1σ	0.018290	0.016611	-9.18%	0.016611	-9.18%
-0.5σ	0.046269	0.045811	-0.99%	0.045811	-0.99%
0σ	0.099477	0.099477	0.00%	0.099477	0.00%
0.5σ	0.185578	0.186709	0.61%	0.186709	0.61%
1σ	0.307511	0.319481	3.89%	0.319481	3.89%
1.5σ	0.463779	0.513493	10.72%	0.462853	-0.20%
2σ	0.651430	0.789272	21.16%	0.651430	0.00 %
2.5σ	0.868926	1.173590	35.06%	0.869707	0.09 %
3σ	1.117138	1.701299	52.29%	1.122943	0.52 %

Table 4.2. Quadratic approximations of a risk profile

Clearly, the accuracy of the approximation increases with the number of supporting points considered. Note that the extremely large error in the case of a $+2.5\sigma$ or $+3\sigma$ drop reflects the fact that the option is deep out-of-the-money and therefore almost valueless. The increase in accuracy is particulary important for large positive market shocks (see Fig. 4.3): a long call has an unlimited profit potential but limited downside risk in the event of a decrease of the risk factor price.

The numerical effort associated with piecewise quadratic approximation is directly related to the number of supporting points considered. In case of a M-dimensional risk factor space, an approximation with n supporting points requires n full valuations, n M-dimensional vectors of the first derivatives and n $M \times M$ Hessian matrices for which $n\frac{(M+1)M}{2}$ items have to be evaluated. Clearly, the number of supporting points should be kept small. Moreover, the

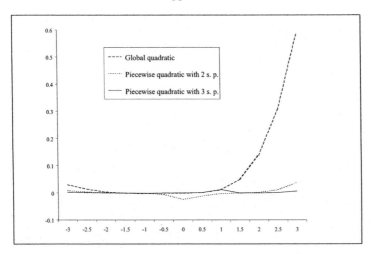

Figure 4.3. Absolute error of the quadratic approximations

locations of the supporting points is crucial for the goodness of the approximation. The choice of the adequate locations, especially for high dimensional
risk factors, is not trivial and will be investigated in the next chapter.

5. Approximation of the Portfolio Distribution

As discussed in Chap. 2, the measurement of market risk requires a complete description of the probabilistic behaviour of the portfolio (or alternatively of the profit and loss function, respectively of the return function) at the end of the holding period. This is achieved by deriving the distribution F_g, ($F_{\bar{g}}$, respectively F_{R_g}) from the distributional information (the distribution of the risk factors) and the price information (the pay-off profile), with:

$$F_g(v) = P\{w|g[w(T)] \leq v\}. \tag{5.1}$$

Considering high-dimensional and non-linear portfolios, an exact representation of F_g may be not available or may involve a considerable analytical effort, which is hardly justified for portfolios with frequent re-balancing.

This chapter discusses various methods for the approximation of the portfolio distribution. The *analytical approximation* methods (see Sect. 5.1) provide an analytical representation of F_g by assuming joint normality of the risk factor distribution and by substituting the pay-off profile with a global Taylor approximation. In case the Taylor approximation is of order one, the approximation of F_g is called Delta approximation. If the Taylor approximation is of order two, the approximation of F_g is referred to as the Delta-Gamma approximation. The joint normality assumption does not allow to capture the observed leptokurtosis of the risk factor distribution and consequently the analytical approximations may give an altered view of the portfolio distribution.

Scenario-based methods are much more flexible since they pose no assumption on the risk factor distribution. These methods involve the generation of a sample of scenarios, i.e. of possible outcomes for the value of the risk factors at time T. Two sampling schemes for the generation of scenarios, the *pseudo-random* and the *quasi-random* method, are presented in Sect. 5.2. A common scenario-based method for the approximation of the portfolio distribution is the *Monte Carlo simulation* which is discussed in Sect. 5.3. Mostly, the Monte Carlo simulation consists of the full valuation of the portfolio for N generated scenarios. The N evaluations of the portfolio represent a random sample from the portfolio distribution. The distribution of the sample, the empirical distribution, is then defined to be the discrete distribution obtained by assigning a probability of $\frac{1}{N}$ to each evaluation of the portfolio and represents the Monte Carlo approximation of F_g. The accuracy of the approximation is directly related to the sample size.

Section 5.4 introduces a new scenario-based methodology, the Barycentric Discretisation with Piecewise Quadratic Approximation (BDPQA), which involves the generation of two sets of scenarios. The first set S_1 with sample size N^1 is equivalent to the scenario set used in the Monte Carlo scheme. The set S_2 with sample size $J << N^1$ is composed of distinguished scenarios, called *generalised barycenters*, which are obtained by a *barycentric discretisation* of the risk factor space. For each scenario in S_1 the risk profile is approximated by the means of a quadratic approximation whose supporting point is an element of S_2. As a result, the BDPQA yields a piecewise quadratic approximation of the risk profile.

Finally, Sect. 5.5 analyses the accuracy of the BDPQA and of the Delta-Gamma approximation for the test portfolios presented in Sect. 1.6.

5.1 Analytical Methods

An exact analytical representation of the portfolio distribution is not possible when the pay-off profile is given numerically and/or the unconditional distribution of the risk factors is not available (as for GARCH models for instance). Even in presence of analytical value functions and existing risk factor distribution, determining the exact portfolio distribution is a complex problem, in particular for portfolios with option content.

Analytical solutions for optioned portfolios with the strong assumptions that the options are held until expiration and that the time to expiration is equal to the holding period can be found in Bookstber and Clarke (1983a), Bookstber and Clarke (1983b), and Brooks (1991). Mostly, the approaches consist in investigating for each option whether it is in-the-money or out-of-the-money at the terminal date T. Considering a three-dimensional portfolio composed of three stocks w_1, w_2, and w_3, of a call option c_1 on w_1 with strike k_1, and of a call c_2 on w_2 with strike k_2, the end of period value of the portfolio, with corresponding portfolio weights $\theta_1 \cdots \theta_5$, is given with :

$$g[w(T)] = \sum_{i=1}^{3} \theta_i w_i(T) + \theta_4 c_1(k_1, w_1(T)) + \theta_5 c_2(k_2, w_2(T)). \qquad (5.2)$$

$F_g(v) = P(g[w(T)] \leq v)$, the probability that the portfolio value is less than some v may be expressed in terms of conditional probabilities (see for instance Bookstber and Clarke (1983b), p.38):

$$
\begin{aligned}
F_g(v) &= P(g[w(T)] \leq v | w_1(T) \leq k_1, w_2(T) \leq k_2) P(w_1(T) \leq k_1, w_2(T) \leq k_2) \\
&+ P(g[w(T)] \leq v | w_1(T) \leq k_1, w_2(T) > k_2) P(w_1(T) \leq k_1, w_2(T) > k_2) \\
&+ P(g[w(T)] \leq v | w_1(T) > k_1, w_2(T) \leq k_2) P(w_1(T) > k_1, w_2(T) \leq k_2) \\
&+ P(g[w(T)] \leq v | w_1(T) > k_1, w_2(T) > k_2) P(w_1(T) > k_1, w_2(T) > k_2)
\end{aligned}
$$
(5.3)

with for instance:

$$
P(g[w(T)] \leq v | w_1(T) \leq k_1, w_2(T) \leq k_2) = \int_{-\infty}^{k_1} \int_{-\infty}^{k_2} \int_{-\infty}^{g(w_1(T), w_2(T))} f_{w_{1,2,3}} \, dw_3 dw_2 dw_1,
$$

$$
P(w_1(T) \leq k_1, w_2(T) \leq k_2) = \int_{-\infty}^{+\infty} \int_{-\infty}^{w_2(T)} \int_{-\infty}^{w_1(T)} f_{w_{1,2,3}} \, dw_1 dw_2 dw_3,
$$
(5.4)

where $g(w_1(T), w_2(T)) = v - \theta_1 w_1(T) - \theta_2 w_2(T)$ and given that $w(T)$ is a continuous random variable with joint density $f_{w_{1,2,3}}$.

For a portfolio with n options, each on a different underlying, the procedure involves the analysis of 2^n combinations and becomes therefore rapidly lengthy with increasing n. Bookstber and Clarke (1983b) propose an approximate algorithm by assuming normally distributed risk factors and by using a single index model as discussed in Sect. 4.2.1, i.e. by assuming that the return on each asset in the portfolio is linearly driven by some index return. The so-called index algorithm is adapted in Pelsser and Vorst (1996) for lognormally distributed risk factors.

An analytical representation of the portfolio distribution can also be obtained by assuming jointly normal distributed risk factors and by substituting the actual pay-off profile with a global Taylor approximation of first order, which yields the Delta approximation, or of second order, which is referred to as the Delta-Gamma approximation. It is important to note, that both the Delta and Delta-Gamma approximations, which will be discussed in the following sections, model the risk factors, and not the logarithmic price changes (log-returns), as jointly normal distributed. As a result, the law of limited liability (see Sect. 3.1.3) may be potentially violated. Considering a one-dimensional risk factor, a first order Taylor expansion of the log-return around the origin gives:

$$\log\left[\frac{w(T)}{w(0)}\right] = \log[w(T)) - \log(w(0)]$$

$$\approx \log[w(0)] + \frac{w(T) - w(0)}{w(T)} - \log[w(0)] \qquad (5.5)$$

$$\approx \frac{w(T) - w(0)}{w(0)}.$$

Hence, if the log-returns are normally distributed (the generic model of Sect. 3.4), the risk factor changes (innovations, market moves) $w(T) - w(0)$, and the risk factors themselves, are approximatively Gaussian. Clearly, the approximation worsens with increasing risk market moves.

5.1.1 Delta Approximation

Global linear approximations g^Δ of the value functions at current price levels:

$$g^\Delta[w(T)] = g[w(0)] + \sum_{m=1}^{M} \frac{\partial g[w(0)]}{\partial w_m}(w_m(T) - w_m(0)), \qquad (5.6)$$

are widespread in classical risk management. Well-known examples are duration analysis in bond management or Delta hedging in portfolio management (see Fabozzi (1996); Gastineau (1992); Hull (2000); Platt (1986)). Equation (5.6) may be rewritten in matrix notation as:

$$g^\Delta[w(T)] = g[w(0)] + \Delta^\top \hat{w}, \qquad (5.7)$$

where $\Delta^\top$ denotes the transpose of the M-dimensional (Gradient) vector of the first order partial derivatives of g at $w(0)$, and $\hat{w}$ the vector of market moves $w(T) - w(0)$.

Substituting the actual risk profile with g^Δ provides analytical ways for determining the portfolio distribution in case the risk factors are assumed as being jointly normal distributed $\mathcal{N}(\mu, \Sigma)$. The distribution of the linear approximation of the profit and loss $\bar{g}^\Delta$ with:

$$\bar{g}^\Delta[w(T)] = \Delta^\top \hat{w}, \qquad (5.8)$$

yields a linear combination of jointly normal distributed random variables and is therefore univariate normally distributed according to $\mathcal{N}(\mu, \Delta^\top \Sigma \Delta)$. Obviously, g^Δ has Gaussian distribution $\mathcal{N}(g[w(0)] + \mu, \Delta^\top \Sigma \Delta)$ and the Delta approximation of F_g, F_g^Δ, is therefore completely described by its mean and its variance. As a result, the estimation of any risk measure is straightforward. For instance, the Value-at-Risk numbers, the respective percentiles of the profit and loss distribution, are given analytically by:

$$v_\alpha^\Delta = z_\alpha \sqrt{\Delta^\top \Sigma \Delta}. \qquad (5.9)$$

The coefficient z_α corresponds to α quantile of the standard normal distribution, i.e. $z_\alpha = 1.64$ for $\alpha = 5\%$, $z_\alpha = 1.88$ for $\alpha = 3\%$, and $z_\alpha = 2.33$ for $\alpha = 1\%$.

As discussed in Sect. 4.5, Taylor approximations of the risk profiles are only accurate in the vicinity of the supporting point. For large market moves, the appropriateness of the Delta approximation depends on the degree of non-linearity of the risk profile, i.e. on the stability of Δ over the range of $w(T)$.

Figure 5.1 displays the one-dimensional risk profile of a short straddle similar to profile PF_3^I, i.e. a combination of short calls and short puts with same strike price. The weighting of the calls and of the puts has been chosen, so that the portfolio is Delta-hedged, which means that the Δ of a portfolio is equal to 0. The value of a Delta-hedged portfolio remains unchanged for small changes in the risk factors. However, large market movements may still have a severe impact on the portfolio value (see the curved line in Fig. 5.1). For such a portfolio the linear approximation of $\bar{g}[w(T)]$ will be zero over the

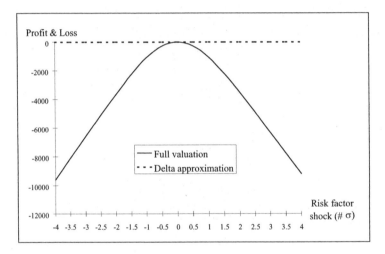

Figure 5.1. Risk profile of short straddle and its Delta approximation

range of possible values of $w(T)$ (the straight line of Fig. 5.1). As a result, the Delta approximation neglects the risk inherent to the portfolio. $F_{g^\Delta} = g[w(0)]$ with probability one and therefore yields a very poor approximation of F_g. As illustrated in Fig. 5.2, the large error stems from the fact that non-linear risk profiles, such as a short straddle, exhibit a large variation of the gradient at Δ over the range of $w(T)$.

The numerical effort related to the Δ approximation is restricted to one full valuation and to the evaluation of the M first order partial derivatives at $w(0)$.

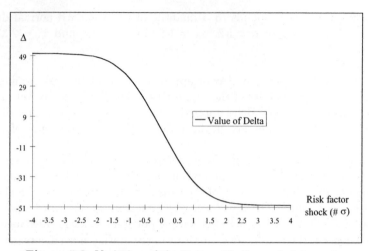

Figure 5.2. Variation of the Delta with the underlying price

5.1.2 Delta-Gamma Approximation

Replacing the linear profile of the Delta approximation by a second order Taylor approximation about the origin:

$$
g^{\Delta-\Gamma}[w(T)] = g[w(0)] + \sum_{m=1}^{M} \frac{\partial g[w(0)]}{\partial w_m} \hat{w}_m
$$
$$
+ \frac{1}{2!} \sum_{m_1=1}^{M} \sum_{m_2=1}^{M} \frac{\partial^2 g[w(0)]}{\partial w_{m_1} \partial w_{m_2}} \hat{w}_{m_1} \hat{w}_{m_2},
\tag{5.10}
$$

which yields in matrix notation:

$$
g^{\Delta-\Gamma}[w(T)] = g[w(0)] + \Delta^\top \hat{w} + \frac{1}{2} \hat{w}^\top \Gamma \hat{w},
\tag{5.11}
$$

allows capturing some non-linearities of the actual profile. $\Gamma \in \mathbb{R}^{M \times M}$ represents the (Hessian) matrix of the second-order derivatives of g at $w(0)$. Note that some entries of Γ may be zero, and therefore $g^{\Delta-\Gamma}[w(T)]$ yields a linear-quadratic profile. Substituting the risk profile with a linear-quadratic function provides additional information on the curvature of g, i.e. on the sensitivities of Δ.

Clearly, the inclusion of quadratic terms transforms the distribution, and the Delta-Gamma approximation of F_g, $F_g^{\Delta-\Gamma}$, is no longer normal. The exact distribution of $g^{\Delta-\Gamma}[w(T)]$, which is derived in Rouvinez (1997) for risk factor changes distributed $\mathcal{N}(0, \Sigma)$, can be found by considering a linear transformation L of the risk factor changes so that the transformed innovations $\hat{w}^*$ are given with:

$$\hat{w}^* = L\hat{w} = (O^\top A^{-1})\hat{w} \tag{5.12}$$

where $\hat{w} \sim \mathcal{N}(0, \Sigma)$, A is the Choleski factorisation of the positive definite covariance matrix Σ, i.e. A is an lower triangular matrix with $AA^\top = \Sigma$, and O is an orthogonal matrix $(O^\top O = I)$ of the eigenvectors of the matrix $A^\top \Gamma A$. Note that $(A^\top \Gamma A)^\top = A^\top \Gamma A$ by construction and therefore $A^\top \Gamma A$ is symmetric. As a result, O always exists and hence L is well defined. Moreover, a diagonalisation of $A^\top \Gamma A$ can be achieved by computing $O^\top (A^\top \Gamma A)O$ (for a discussion see for instance Fraleigh and Beauregard (1995), p.354).

The transformed risk factor changes are jointly normal distributed with covariance matrix Σ^*:

$$\begin{aligned}
\Sigma^* &= L\Sigma L^\top \\
&= O^\top A^{-1} \Sigma (A^\top)^{-1} O \\
&= O^\top A^{-1} AA^\top (A^\top)^{-1} O \\
&= I
\end{aligned} \tag{5.13}$$

and hence independent. By defining:

$$\Delta^* = (L^\top)^{-1}\Delta = (O^\top A^\top)\Delta, \tag{5.14}$$

and:

$$\Gamma^* = (L^\top)^{-1}\Gamma L^{-1} = O^\top (A^\top \Gamma A)O, \tag{5.15}$$

it can be easily verified that:

$$\begin{aligned}
\Delta^{*\top}\hat{w}^* &= \Delta^\top L^{-1} L\hat{w} \\
&= \Delta^\top \hat{w},
\end{aligned} \tag{5.16}$$

and:

$$\begin{aligned}
\hat{w}^{*\top}\Gamma^*\hat{w}^* &= \hat{w}^\top L^\top \left((L^\top)^{-1}\Gamma L^{-1}\right)L\hat{w} \\
&= \hat{w}^\top \Gamma \hat{w}.
\end{aligned} \tag{5.17}$$

Hence, the transformed profit and loss profile $\bar{g}^{*\Delta - \Gamma}$ with:

$$\bar{g}^{*\Delta - \Gamma} = \Delta^{*\top}\hat{w}^* + \frac{1}{2}\hat{w}^{*\top}\Gamma^*\hat{w}^*, \tag{5.18}$$

is equivalent to the linear-quadratic approximation of the profit and loss $\bar{g}^{\Delta - \Gamma} = \Delta^\top \hat{w} + \frac{1}{2}\hat{w}^\top \Gamma \hat{w}$.

Since Γ^* is diagonal, (5.18) may be rewritten as:

$$\bar{g}^{*\Delta-\Gamma} = \sum_{m=1}^{M} \left(\Delta_m^* \hat{w}_m^* + \frac{1}{2} \Gamma_{m,m}^* (\hat{w}_m^*)^2 \right), \qquad (5.19)$$

i.e. as a sum of independent risk factor changes. Reorganising the terms of (5.19) yields:

$$\bar{g}^{*\Delta-\Gamma} = \sum_m \left(\frac{1}{2} \Gamma_{m,m}^* (\hat{w}_m^*)^2 + \Delta_m^* \hat{w}_m^* + \frac{1}{2} \frac{(\Delta_m^*)^2}{\Gamma_{m,m}^*} - \frac{1}{2} \frac{(\Delta_m^*)^2}{\Gamma_{m,m}^*} \right) + \overline{\sum}_m \Delta_m^* \hat{w}_m^*$$

$$= \sum_m \frac{1}{2} \Gamma_{m,m}^* \left(\hat{w}_m^* + \frac{\Delta_m^*}{\Gamma_{m,m}^*} \right)^2 + \overline{\sum}_m \Delta_m^* \hat{w}_m^* - \sum_m \frac{1}{2} \frac{(\Delta_m^*)^2}{\Gamma_{m,m}^*},$$

$$(5.20)$$

where $\sum_m$ denotes the sum over the m's with $\Gamma_{m,m} \neq 0$ and $\overline{\sum}_m$ stands for the sum over the m's with $\Gamma_{m,m} = 0$. Note that since the $\hat{w}_m^*$ are standard normally distributed, the terms $\left(\hat{w}_m^* + \Delta_m^*/\Gamma_{m,m}^* \right)^2$ are distributed $\chi^2_{\{1, \Delta_m^*/\Gamma_{m,m}^*\}}$, i.e. non-central chi-squared with one degree of freedom and non-centrality parameters $\left(\Delta_m^*/\Gamma_{m,m}^* \right)^2$ (see for instance Johnson and Kotz (1970), p.130).

Equation (5.20) may be rewritten as:

$$\bar{g}^{*\Delta-\Gamma} = \sum_m (a_m \, \xi_m) + \xi_0 + c_m, \qquad (5.21)$$

where:

$$a_m = \frac{1}{2} \Gamma_{m,m}^*,$$

$$c_m = -\frac{1}{2} \sum_m \frac{(\Delta_m^*)^2}{\Gamma_{m,m}},$$

$$\xi_m \sim \chi^2_{\{1, \Delta_m^*/\Gamma_{m,m}^*\}},$$

$$\xi_0 \sim \mathcal{N} \left(0, \overline{\sum}_m (\Delta_m^* \hat{w}_m^*)^2 \right),$$

since ξ_0 represents the sum of IID random variables with standard normal distribution. Hence, the distribution $F_{\bar{g}}^{\Delta-\Gamma}$ can be represented as a linear combination of independent, normal and non-central χ^2, random variables. To recover the exact distribution of $F_{\bar{g}}^{\Delta-\Gamma}$ Rouvinez (1997) uses the fact that every distribution is completely determined by a unique characteristic function (see appendix A.2.2) $\phi_{\bar{g}^{\Delta-\Gamma}}(s) = E\left[\exp(is\bar{g}^{\Delta-\Gamma}) \right]$, with $i = \sqrt{-1}$, and that the characteristic function of a linear combination of independent random variables corresponds to the product of their characteristic functions given with (see Feller (1966), p.503 and Johnson and Kotz (1970), p.134):

$$\phi_{\xi_m}(s) = \frac{1}{\sqrt{1 - i\Gamma^*_{m,m}s}} \exp\left[\frac{i\Delta^*_m s}{2\Gamma^*_{m,m}(1 - i\Gamma^*_{m,m}s)}\right],$$

$$\phi_{\xi_0}(s) = \exp\left[-\frac{\overline{\sum_m}(\Delta^*_m)^2 s^2}{2}\right], \tag{5.22}$$

$$\phi_{c_m}(s) = \exp[ic_m s].$$

The characteristic function of $\bar{g}^{\Delta-\Gamma}$ is therefore:

$$\phi_{\bar{g}^{\Delta-\Gamma}} = \exp\left[ic_m s - \frac{\overline{\sum_m}(\Delta^*_m)^2 s^2}{2}\right] \prod_m \frac{\exp\left[\frac{i\Delta^*_m s}{2\Gamma^*_{m,m}(1 - i\Gamma^*_{m,m}s)}\right]}{\sqrt{1 - i\Gamma^*_{m,m}s}}, \tag{5.23}$$

where $\prod_m$ denotes the product over the m's with $\Gamma_{m,m} \neq 0$. Provided that $E\left[|\,g^{\Delta-\Gamma}\,|\right] < \infty$ and that there exists some c and δ so that, for all $s > 1$,

$$|\,\phi_{\bar{g}^{\Delta-\Gamma}}(s)\,| < cs^{-\delta},$$

the distribution function of the profit and loss satisfies (see Gil-Pelaez (1951) and Davies (1973)):

$$F_{\bar{g}^{\Delta-\Gamma}}(x) = \frac{1}{2} - \int_{-\infty}^{\infty} I_v\left(\frac{\phi_{\bar{g}^{\Delta-\Gamma}}(s)\exp[-isx]}{2\pi s}\right)ds, \tag{5.24}$$

where $I_v(x)$ is the modified Bessel function of first order (see for instance Springer (1979), p.419):

$$I_v(x) = \sum_{i=0}^{\infty} \frac{(x/2)^{2i+v}}{i!(i+v)!}.$$

Equation (5.24) may be rewritten:

$$F_{\bar{g}^{\Delta-\Gamma}}(x) = \frac{1}{2} + \frac{1}{2\pi} \int_0^{\infty} \frac{\exp(ixs)\phi_{\bar{g}^{\Delta-\Gamma}}(-s) - \exp(-ixs)\phi_{\bar{g}^{\Delta-\Gamma}}(s)}{is} ds. \tag{5.25}$$

The α-quantile of the profit and loss distribution (VaR$_\alpha$) is obtained by the inversion of $F_{\bar{g}^{\Delta-\Gamma}}$:

$$-\text{VaR}_\alpha = F^{-1}_{\bar{g}^{\Delta-\Gamma}}(\alpha), \tag{5.26}$$

and therefore requires the inversion of $\phi_{\bar{g}^{\Delta-\Gamma}}$. The corresponding algorithms, based on Imhoff (1961)'s numerical inversion technique, can be found in Davies (1973, 1980).

Figure 5.3 displays the risk profile of the Delta-hedged short straddle and compares the Delta-Gamma approximation with full valuation. Clearly, the

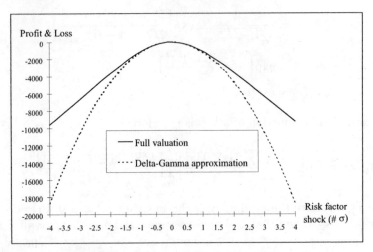

Figure 5.3. Risk profile of short straddle and its Gamma approximation

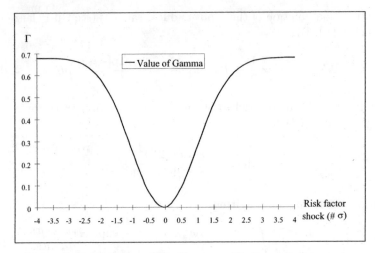

Figure 5.4. Variation of the Gamma with the underlying price

quadratic approximation provides increased accuracy compared to the Delta approximation. However, for shocks larger than $\pm 2\sigma$ the Delta-Gamma approximation tends to overestimate the risk. The error is due to the variability of Γ over the range of $w(T)$ (see Fig. 5.4). The numerical effort for determining quadratic approximations of g increases approximatively with order 2 in the dimension M. This is due to evaluating $\frac{M(M+1)}{2}$ items of Γ.

5.2 Generation of Scenarios

Scenario-based methods typically analyse the behaviour of the portfolio of interest for realisations of the risk factors at time T. Since it is clearly impractical to consider all feasible scenarios, a sample of possible values $w^i, i = 1, \cdots N$ for $w(T)$ has to be drawn from the risk factor space. This section discusses the generation of scenarios, which is used for both the Monte Carlo Simulation and the BDPQA. Note that the methodology discussed in the sequel is not equivalent to the *scenario analysis*, which consists mainly in selecting a small set of "representative" scenarios for which the associated probabilities are assigned subjectively.

The two scenario generation methods presented in this chapter, the pseudo-random (Sect. 5.2.1) and the quasi-random schemes (Sect. 5.2.2) do not choose scenarios arbitrary but proceed by sampling from the distribution that describes the dynamics of the risk factors. This requires an effective way of generating numbers uniformly distributed on the unit interval $[0, 1]$, in short uniform deviates or $U(0, 1)$. Specific transformations from uniform deviates, which will be presented in Sect. 5.2.3, allow for generating sequences of numbers with arbitrary continuous distributions.

Both methods are purely deterministic and generate a sequence of numbers the properties of which are only similar to those of a sequence of uniform deviates. These properties are mainly the equidistribution in the interval $[0, 1)$ measured by the discrepancy (see Sect. 5.2.3) and statistical independence. The pseudo-random method, the classical scenario-generating method, tends to mimic a sequence of independent and uniformly distributed random variables. The quasi-random method focusses on the equidistribution property and provides algorithms to fill the interval $[0, 1)$ more evenly than the pseudo-random scheme.

Both the pseudo- and the quasi-random generation of scenarios rely on the appropriateness of the modelling of the risk factor dynamics. An alternative approach, called *historical simulation*, uses as scenarios a sequence of past observations of the risk factors. The historical simulation attributes equal weight to each observation. Hence, the historical simulation avoids the modelling of the risks factors and therefore does not make explicit assumption about their dynamics. Moreover, the method captures actual departures from normality and correlations between the risk factors. The underlying assumption of the historical simulation is that the sequence of observations at disposal represents a representative sample of the risk factor distribution. This is only true if the stochastic process governing the data is stationary. As it will be discussed below, scenarios-generated methods require a large set of scenarios to achieve sufficient accuracy. To produce 4'000 scenarios the historical simulation needs a record of about 16 years of daily observations.

It is more than questionable that the stochastic process remains stationary over a so long period of time.

5.2.1 The Pseudo-Random Method

Pseudo-random generators aim to produce large sequences of statistically independent numbers which are distributed as $U(0,1)$. Almost all pseudo-random generators used in practice belong to the class of *linear congruential generators* introduced by Lehmer (1951) and yield following recurrence (see for instance Ripley (1987), p.20):

$$x^n = (ax^{n-1} + c) \text{ modulo } m_o, \qquad (5.27)$$

where the *multiplier* a, the *shift* c, and the *modulus* m_o are all non-negative integers. Generators for which $c = 0$ are called *multiplicative*. The sequence has to be initialised by a starting value x^0 called the *seed*. Pseudo-random numbers u^n on [0,1) are then obtained by the transformation:

$$u^n = \frac{u^n}{m_o} \qquad (5.28)$$

From (5.27), it is easy to verify that the sequence recurses as soon as a number is repeated and is therefore *periodic* with a period of length p. Moreover, $m_o + 1$ values $(x^0, \cdots x^{m_o})$ can not be distinct. As a result, $p \leq m_o$. It can be shown that a congruential generator has full period, i.e. $p = m_o$ if, and only if, (see Knuth (1998), p.17):

- the greatest common divisor of c and m_o is one, i.e. c is relatively prime to m_o,
- $a = 1$ modulo f for each prime factor f of m_o,
- $a = 1$ modulo 4 if m_o is a multiple of 4.

For multiplicative generators the maximum period is $m_o - 1$ since the sequence is repeated indefinitely if $x^n = 0$ occurs. A multiplicative generator has maximum period if, and only if, a is a primitive root of a prime number m_o, i.e. (see Fishman (1996) pp.592-3):

- $a^{m_o-1} - 1 = 0$ (modulo m_o),
- for all integers $i < m_o - 1$, the quantity $\frac{a^i - 1}{m_o}$ is not an integer.

The generated sequence should exhibit a sufficiently large period length p so that it never cycles for practical applications. Therefore a, c, and m_o should be chosen in a way as to produce a period of maximal length and m_o should be as large as possible. For technical reasons, i.e. to avoid the use of multi-precision arithmetic, m_o is generally bounded above by the computer word size, which amounts 2^b for a b-bit binary computer and allows the representation of $2^b - 1$ non-negative integers. The period length is an important but clearly

insufficient criterion to assess the adequacy of a generator. It is easy to verify
that following recurrence:

$$x^n = (x^{n-1} + 1) \text{ modulo } m_o,$$

has a period of length m_o but is completely predictable. Further insight in
the choice of the multiplier, of the shift, and of the modulus is provided
by randomness tests. As stated above, the objective of the pseudo-random
scheme is to mimic a sequence of uniform deviates. The requirements imposed
on the generator refer mostly to *equidistribution* and *statistical independence*
properties. Both properties are analysed by a series of theoretical and empirical tests. *Theoretical tests* analyse the mathematical structure of generators
by using number-theoretic arguments. They are very powerful since they assess a priori the appropriateness of the generators analysed but are generally
difficult to construct. *Empirical* tests assess a posteriori the randomness properties of sequences of numbers produced by the generators of interest. Note
that linear generators are periodic, i.e. can generate only a finite set of values
and therefore possess a certain structure. Mathematically, the N generated
numbers always lie on a finite number of parallel hyperplanes (for a definition of hyperplane, see Sect. 5.4.1) in $(0,1)^N$ and therefore exhibit a lattice
structure which is not conform with true randomness. Un upper bound on
the maximal number of hyperplanes is $(n!/m_o)^{1/n}$, provided that m_o is prime
(see Marsaglia (1968)).

Theoretical tests of equidistribution include the computation of discrepancy, the calculation of the number of parallel hyperplanes (which should be
as high as possible) and of the distance between adjacent hyperplanes (which
should be as short as possible). Equidistribution properties of a generated
sequence may be analysed empirically by the Kolmogorov-Smirnov or frequency test (see Sect. 5.3.1). Empirical tests of independence are for instance
serial tests, the estimation of the serial correlation of the sequence, and permutation tests. A detailed description of theoretical as well as empirical tests
of randomness can be found in Knuth (1998), pp.41-118 and Fishman (1996),
pp.611-634.

5.2.2 The Quasi-Random Method

For many applications, such as the numerical integration in high dimensions,
the value of the scenario generating method is not primarily the randomness
of the sampling but the equidistribution properties in the risk factor space.
The primary objective of the quasi-random method is to generate more evenly
distributed points in the risk factor space, *low-discrepancy sequences*, than the
pseudo-random method does. The construction of quasi-random sequences
relies on number theory whose theoretical underpinnings can be found in
Niederreiter (1992). A standard measure of equidistribution in a given volume

is the *discrepancy*, which is generally defined over the M-dimensional unit cube I^M. Given a set of points $u^1, u^2, \cdots u^N \in I^M$, a subset $G \in I^M$, and its associated counting function $C(G)$, i.e. the function indicating the number of points $u^n \in G$:

$$C(G) = \sum_{n=1}^{N} I_{\{u^n \in G\}} \quad (I \text{ is the indicator function}), \qquad (5.29)$$

the discrepancy D_N of the point set is (see Niederreiter (1992), p.14):

$$D_N(\mathcal{G}; u^1, \cdots, u^N) = \sup_{G \in \mathcal{G}} \left| \frac{C(G)}{N} - \lambda^M(G) \right|, \qquad (5.30)$$

where $\mathcal{G}$ is a (non-empty) family of Lebesgue-measurable subsets of I^M and λ^M the M-dimensional Lebesgue measure (see appendix A.1). Hence, the discrepancy is defined as the supremum of the absolute difference between the number of points in a subset of I^M divided by N and the volume (more precisely the M-dimensional Lebesgue-measure) of the subset considered. Intuitively, the more evenly a set of points is spaced over I^M the lower is its discrepancy.

The important special cases of $D_N(\mathcal{G}; u^1, \cdots, u^N)$, the extreme and the star discrepancy, are used in practice to analyse the equidistribution property of a point set:

- The *extreme discrepancy* $D_N(u^1, \cdots, u^N) = D_N(\mathcal{J}; u^1, \cdots, u^N)$, where $\mathcal{J}$ is the family of interval of I^M of the from $\prod_{m=1}^{M} [u_m, u_m)$, i.e. the extreme discrepancy is defined over all rectangular M-dimensional regions of I^M.
- The *star discrepancy* $D_N^*(u^1, \cdots, u^N) = D_N(\mathcal{J}^*; u^1, \cdots, u^N)$, where $\mathcal{J}^*$ is the family of interval of I^M of the from $\prod_{m=1}^{M} [0, u_m)$, i.e. the star discrepancy is defined over all rectangular M-dimensional regions with one vertex at the origin.

Note that the extreme and star discrepancies are directly related by following inequalities: (see Niederreiter (1992), p.15):

$$D_N^*(u^1, \cdots, u^N) \leq D_N(u^1, \cdots, u^N) \leq 2^M D_N^*(u^1, \cdots, u^N).$$

The quasi-random method chooses sequences $\{u^n\}$ so as to minimise D_N, respectively D_N^*, i.e. tends to minimise the difference between the proportion of points in each rectangular subset of I^M and its measure. Sequences for which the order of magnitude of the discrepancy is not greater than $N^{-1}(\log N)^M$ are generally referred to as low-discrepancy sequences. Classical low-discrepancy sequences are proposed by Halton (1960), Sobol (1976, 1979), and Faure (1982). For these three sequences the star discrepancy satisfies (see Fox (1986)):

$$D_N^*(u^1, \cdots, u^N) = K_M \frac{(\log N)^M}{N} + O\left(\frac{(\log N)^{M-1}}{N}\right), \qquad (5.31)$$

where K_M is a constant, the value of which is dependent on the number of dimensions M and on the sequence considered. Note that the star discrepancy of uniform deviates is of order $\sqrt{\frac{\log \log N}{N}}$ and hence decreases much slower than its low-discrepancy counterpart with increasing N. However, contrary to its quasi-random counterpart, the star discrepancy of a random sequence does not depend on M. It is conjectured, that for any *sequence* (see Niederreiter (1992), p.33):

$$D_N^*(u^1, \cdots, u^N) \geq K\left(\frac{(\log N)^M}{N}\right), \qquad (5.32)$$

for some constant K, holds. This has been shown for $M = 1$ and $M = 2$. Note that finite points *sets*, such as the *Hammersley* point sets, may exhibit a discrepancy improved by one logarithmic term (see Niederreiter (1992), p.33):

$$D_N^*(u^1, \cdots, u^N) \geq K\left(\frac{(\log N)^{M-1}}{N}\right). \qquad (5.33)$$

However, infinite sequences have the advantage over point sets that the number N of points generated do not have to be specified in advance.

The K_M values of the *Halton, Sobol, and Faure* sequences for selected M are displayed in table 5.1 and indicate that Faure's algorithm is superior to the others with respect to the order of magnitude of D_N^*. Several empirical studies (see for instance Fox (1986); Bratley and Fox (1988); Paskov (1997); Galanti and Jung (1997), and Berman (1998)) show that, for numerical integration purposes, the Faure and Sobol constructions outperform the Halton's algorithm. The numerical results reported in Galanti and Jung (1997); Boyle et al. (1997) suggest that the Sobol algorithm performs better than the Faure construction, especially for high-dimensional problems.

♯ dim (M)	3	4	5	6	7	8	13
K_M: Halton	0.81	1.25	2.62	6.13	17.3	52.9	90580
K_M: Sobol	1.00	1.44	1.66	3.20	5.28	15.2	647
K_M: Faure	0.12	0.099	0.024	0.018	0.0041	0.0088	0.00001

Table 5.1. Star discrepancy: constant K_M for low-discrepancy sequences

The construction of the three sequences is based on number theory and relies on the *expansion* of non-negative integers in an arbitrary base b, given b is prime and ≥ 2. The base b number of the integers is then transformed into a number in the interval $(0, 1)$ by *reflection about the decimal point* (for an excellent discussion, see Galanti and Jung (1997). A simple example with $b = 2$ and $n = 5$ illustrates the procedure. The expansion of 6 in base 2 is given with:

$$6 = \mathbf{1}(2^2) + \mathbf{1}(2^1) + \mathbf{0}(2^0).$$

The coefficients of the expansion are therefore "110". The reflection about the decimal point is then obtained by reversing the digits of "110", which yields "011" and by putting a so-called radix point, i.e. a decimal point in base b, which gives "0.011". The base 2 number "0.011" has decimal representation:

$$\text{"0.011"} = 0\frac{1}{2^1} + 1\frac{1}{2^2} + 1\frac{1}{2^3} = \frac{3}{8}.$$

Formally, let $b \geq 2$ be an integer and $Z_b = \{0, 1, .., b-1\}$ the residual modulo b. Any integer $n \geq 0$ has a unique *digit expansion* (see Niederreiter (1992), p.24)

$$n = \sum_{j=0}^{\infty} a_j(n) b^j \tag{5.34}$$

in term of base b. The expansion is referred to as the *b-adic expansion* of n. $a_j(n)$ is an element of Z_b and equals zero for all sufficiently large j, which ensures that the b-adic expansion is always finite. The *radical inverse function* ϕ_b, the extension about the decimal point, is given with:

$$\phi_b(n) = \sum_{j=0}^{\infty} a_j(n) b^{-j-1}. \tag{5.35}$$

The infinite one-dimensional *van der Corput sequence* $u^0, u^1, \cdots$ is obtained by setting $u^n = \phi_b(n)$ for all $n > 0$. A M-dimensional van der Corput sequence where

$$u^n = (\phi_{b_1}(n), \cdots, \phi_{b_M}(n)), \ b_1, \cdots, b_M \in \mathbb{N} \geq 2, \ b_i \neq b_j \text{ for } i \neq j$$

is referred to as a *Halton sequence*. Hence the Halton sequence is obtained by computing the radical inverse function for M different bases b_m. The use of a different base for each m ensures that the sequence is not identical in each dimension.

Unlike the Halton sequence, the *Faure construction* considers the same base b for each dimension, b is the smallest prime number that is greater than or equal to the number of dimension M. In order to avoid the sequence being identical in all dimensions, the numbers within the mth dimension, $m > 1$, are computed as follows (see Galanti and Jung (1997)):

$$u^n = \sum_{i=0}^{\infty} f_i(a_j(n))b^{-i-1}, \tag{5.36}$$

where:

$$f_i(a_j(n)) = \left[\sum_{j \geq i}^{\infty} (m-1)^{j-1} \binom{j}{i} a_j(n) \right] \text{ modulo } b. \tag{5.37}$$

$\binom{j}{i}$ denotes the binomial coefficients $\frac{j!}{i!(j-i)!}$. Clearly, $f_i(a_j(n))$ yields a permutation of the $a_j(n)$ and causes therefore a different ordering of the digit expansion for each dimension.

The *Sobol sequence* is, similarly to the Faure sequence, essentially a reordering of the Halton sequence. Unlike the Faure sequence, the Sobol sequence uses, independently of the number of dimensions, always binary expansions, i.e. b equals two.

The one-dimensional Sobol sequence $u^1, u^2, \cdots$ has following representation:

$$u^n = b_1 v_1 \oplus b_2 v_2 \cdots \tag{5.38}$$

where $\cdots b_2 b_1$ is the binary representation of n, the v_i are *direction numbers*, and $\oplus$ denotes the bit-by-bit exclusive OR operator (XOR), i.e. $b_1 v_1 \oplus b_2 v_2$ equals 0 if both sides of $\oplus$ are 1 or 0 and equals 1 otherwise. The direction numbers are obtained by the following d-term recursion formula:

$$v_i = a_1 v_{i-1} \oplus a_2 v_{i-2} \oplus \cdots \oplus a_d v_{i-d} \oplus \frac{v_{i-d}}{2^d}, i > d, \tag{5.39}$$

where the $a_i \in \{0, 1\}$ are the coefficients of an arbitrarily chosen primitive polynomial $p(z)$ of order d:

$$p(z) = z^d + a_1 z^{d-1} + \cdots + a_{d-1} z + 1, \tag{5.40}$$

A list of primitive polynomials can be found for instance in Press et al. (1992), p.311. By setting $m_i = \frac{v_i}{2^i}$, recurrence (5.39) can be rewritten:

$$m_i = 2a_1 m_{i-1} \oplus 2^2 a_2 m_{i-2} \oplus \cdots \oplus 2^d a_d m_{i-d} \oplus m_{i-d}, i > d \tag{5.41}$$

The initial values $m_i, i = 1, \cdots, d$ are arbitrary odd integers provided that $m_i < 2^i$. Note that the efficient algorithm of Antonov and Saleev (1979), using the *Gray code representation* of n is generally implemented instead of Sobol's original method (for a discussion, see Bratley and Fox (1988) and Press et al. (1992), pp.310-311). A M-dimensional Sobol sequence is obtained by constructing M one-dimensional sequences, each based on a different primitive polynomial. Although the choice of the initial values has asymptotically no impact on the uniformity properties of the sequence, it may be of importance, in the M-dimensional case, if only small sequences are considered. Mostly, it is preferable to chose different m_i for each dimension. Sobol (1976) provides criteria for the suitable choice of the initial values.

To illustrate how low-discrepancy sequences behave, a one-dimensional Sobol scheme with following primitive polynomial of degree $d = 3$:

$$p(z) = z^3 + z^2 + 1 \tag{5.42}$$

will be constructed. The recurrence (5.41) becomes:

$$\begin{aligned} m_i &= 1(2m_{i-1}) \oplus 0(2^2 m_{i-2}) \oplus 1(2^3 m_{i-3}) \oplus m_{i-3} \\ &= 2m_{i-1} \oplus 8m_{i-3} \oplus m_{i-3} \end{aligned} \tag{5.43}$$

Setting $m_1 = 1, m_2 = 3$ and $m_3 = 3$, the $m_i, i > 3$ are obtained iteratively:

$$\begin{aligned} m_4 &= 6 \oplus 8 \oplus 1 \\ &= 0110 \oplus 1000 \oplus 0001 \qquad \text{in binary} \\ &= 1111 \qquad\qquad\qquad\quad \text{in binary} \\ &= 15 \end{aligned}$$

$$\begin{aligned} m_5 &= 30 \oplus 24 \oplus 3 \\ &= 11110 \oplus 11000 \oplus 00011 \qquad \text{in binary} \\ &= 100101 \qquad\qquad\qquad\qquad \text{in binary} \\ &= 5 \end{aligned}$$

and so one. The corresponding direction numbers $v_i = \frac{m_i}{2^i}$ are displayed in

i	1	2	3	4	5
v_i	$\frac{1}{2}$	$\frac{3}{4}$	$\frac{3}{8}$	$\frac{15}{16}$	$\frac{5}{32}$
v_i in binary	0.1	0.11	0.0011	0.1111	0.00101

Table 5.2. Direction numbers of the Sobol sequence

table 5.2. The Sobol sequence may now be generated by using formula (5.38):

$$n = 1 : b_1 = 1$$

$$u^1 = v_1 = \frac{8}{16}$$

$$n = 2 : b_2 = 1, \ b_1 = 0 \ (\text{since } 2 = \text{"10" in binary notation})$$

$$u^2 = v_2 = \frac{12}{16}$$

$$n = 3 : b_2 = 1, \ b_1 = 1$$

$$u^3 = v_1 \oplus v_2$$

$$= 0.1 \oplus 0.11 \qquad \text{in binary}$$

$$= 0.01 \qquad\qquad \text{in binary}$$

$$= \frac{4}{16}$$

(5.44)

etc.

The first 15 numbers of the Sobol sequence are:

$$\frac{8}{16}, \frac{12}{16}, \frac{4}{16}, \frac{6}{16}, \frac{14}{16}, \frac{10}{16}, \frac{2}{16}, \frac{15}{16}, \frac{7}{16}, \frac{3}{16}, \frac{11}{16}, \frac{9}{16}, \frac{1}{16}, \frac{5}{16}, \frac{13}{16}$$

and it is easy to observe in Fig. 5.5 that each new number added to the sequence tends to fill in the gaps in the existing sequence. Figure 5.6 displays the first and second dimensions of 1'024 52-dimensional points generated by the pseudo-random method (left) and by the Sobol algorithm (right). Clearly, the Sobol sequence fills the two-dimensional space in a more regular way than its pseudo-random counterpart, which exhibits clusters of points. Low-discrepancy sequences are constructed in a way that the numbers in a sequence "avoid" each other. This kind of constructions generates correlations between the points and the sequences may no longer be considered as sets of independent deviates.

Note also that the uniformity properties of low-discrepancy sequences deteriorate with increasing M, and the quasi-random scheme may suffer in high dimensions from a lattice structure that is more pronounced than the pseudo-random algorithm. Figure 5.7 shows the 51th and 52th dimension of both sequences. The Sobol sequence displays patterns and yields a very poor coverage of the unit square.

5.2.3 Generation of Distributions for the Risk Factors

The quasi- or pseudo-random generated sequence of numbers $u^n, n = 1, \cdots N$ distributed uniformly on (0,1) has to be transformed into a sequence of numbers w^n with M-dimensional distribution that describes the risk factor dynamics. The generation of M-dimensional random vectors depends on the

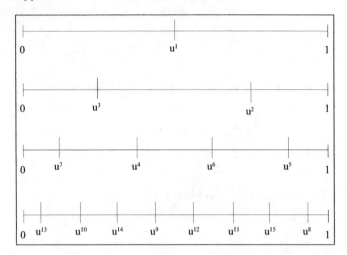

Figure 5.5. One-dimensional Sobol sequence

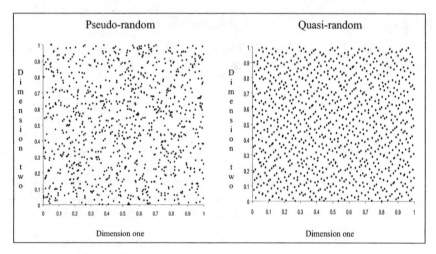

Figure 5.6. Pseudo- and quasi-random uniform deviates: dim 1 and 2

ability to generate univariate random variables. The *inverse transform algorithm* represents a simple method for the generation of univariate random variables from a sequence of uniform deviates (see for instance Ross (1990), p.60).

Let's consider an arbitrary continuous distribution function F and a random variable $X = F^{-1}(U)$, where U is uniform on (0,1). It is not difficult to verify that the distribution function $F_X(x)$ of X equals $F(x)$:

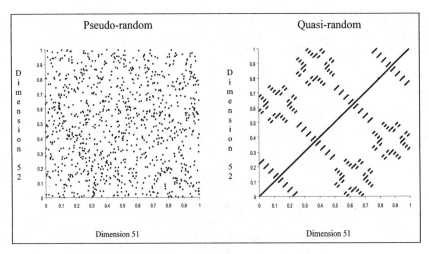

Figure 5.7. Pseudo- and quasi-random uniform deviates: dim 51 and 52

$$F_X(X) = P(X \leq x)$$
$$= P\big(F^{-1}(U)\big) \leq x$$
$$= P(F(F^{-1}(U)) \leq F(x)$$

since $F(x)$ is strictly increasing   (5.45)

$$= P(U \leq F(x))$$
$$= F(X)$$

since U is uniformly distributed on $(0,1)$.

Hence, the generation of random variables with a continuous distribution function F is obtained by generating uniform deviates u^n as discussed in the preceding sections and by computing $w^n = F^{-1}(u^n)$. Unfortunately, the inversion of F is generally not given analytically and requires iteration procedures which may be very slow. Moro (1995) provides a fast algorithm for the inverse normal function, the maximal error of which is $3 \cdot 10^{-9}$ for all values of the cumulative normal distribution function which lie in the interval $[10^{-12}, 1 - 10^{-12}]$.

Recall that all models for the log-returns which have been presented in Chap. 3 are conditionally or unconditionally normal or Student-t distributed. Transformation methods, which allow the generation of Gaussian and Student-t random variables from uniform deviates by obviating the computation of F^{-1}, are discussed in the sequel. Note, however, that these algorithms, based on a transformation into polar coordinates, alter the order of the generated sequence and should not be used in conjunction with the low-discrepancy scheme (see for instance Galanti and Jung (1997)).

The Normal Distribution Suppose that X and Y are independent standard normal random variables. Their joint density f_{XY} is given with (see for instance Rice (1995), p.96):

$$f_{XY}(x,y) = \frac{1}{2\pi}e^{-\left(\frac{x^2}{2}-y^2 2\right)}.$$

Let R and Θ denote the polar coordinates of the vector (X,Y):

$$\begin{aligned} X &= R\cos\Theta, \\ Y &= R\sin\Theta, \end{aligned} \tag{5.46}$$

and J be the Jacobian of the transformation:

$$\begin{aligned} J &= \det \begin{vmatrix} \frac{\partial r}{\partial x} & \frac{\partial r}{\partial y} \\ \frac{\partial \theta}{\partial x} & \frac{\partial \theta}{\partial y} \end{vmatrix} \\ &= \det \begin{vmatrix} \frac{x}{\sqrt{x^2+y^2}} & \frac{y}{\sqrt{x^2+y^2}} \\ \frac{-y}{x^2+y^2} & \frac{x}{x^2+y^2} \end{vmatrix} \\ &= \frac{1}{r}. \end{aligned} \tag{5.47}$$

The joint density of R and Θ is therefore:

$$\begin{aligned} f_{R\Theta}(r,\theta) &= \frac{1}{J}f_{XY}(r\cos\theta, r\sin\theta) \\ &= \frac{r}{2\pi}e^{\left(\frac{-r^2\cos^2\theta}{2}-\frac{r^2\sin^2\theta}{2}\right)} \\ &= \frac{1}{2\pi}re^{\frac{-r^2}{2}}, \end{aligned} \tag{5.48}$$

and it results that R and Θ are independent. Θ is uniformly distributed with density $\frac{1}{2\pi}$ on $[0,2\pi]$ and R has Rayleigh density $re^{\frac{r^2}{2}}$ on $[0,1]$. Defining $S = R^2$, it is easy to verify that:

$$F_S(s) = P(S \leq s) = P(R \leq \sqrt{s}) = F_R(\sqrt{s}), \tag{5.49}$$

and therefore:

$$f_S(s) = \frac{d}{ds}F_R(\sqrt{s}) = \frac{1}{2}t^{-\frac{1}{2}}f_R(\sqrt{s}) = \frac{1}{2}s^{-\frac{1}{2}}s^{\frac{1}{2}}e^{-\frac{1}{2}} = \frac{1}{2}e^{-\frac{1}{2}}. \tag{5.50}$$

Hence, S is exponentially distributed with parameter $\frac{1}{2}$ and can be obtained by generating an uniformly distributed variable U_1 on $[0,1]$ and setting:

$$S = \sqrt{-2\log U_1}. \tag{5.51}$$

The generation of a pair of independent normal random variables X and Y can be therefore accomplished as follows:

- Generate U_1 and U_2 which are independent and uniform on $[0,1]$.
- Compute $-2\log U_1$ which is exponential with parameter $\frac{1}{2}$.
- Compute $2\pi U_2$ which is uniform on $[0,2\pi]$.
- Return $X = \sqrt{-2\log U_1}\cos(2\pi U_2)$ and $Y = \sqrt{-2\log U_1}\sin(2\pi U_2)$ which are independent standard normal.

This method is called the method of *Box-Muller*. The need to compute trigonometric functions is not very efficient. This can be avoided by using a slightly different algorithm called the *polar method*. The polar method requires the generation of uniformly distributed variables in the circle of radius one and centered at the origin. This can be easily achieved by (see for instance Ross (1990), p.71):

- generating U_1 and U_2 which are independent and uniform on $[0,1]$,
- computing $V_1 = 2U_1 - 1$ and $V_2 = 2U_1 - 2$,

  The pair (V_1, V_2) is therefore uniformly distributed in the square centered at the origin with side length = 2.
- accepting (V_1, V_2) only if $V_1^2 + V_2^2 \leq 1$.

  It follows that (V_1, V_2) is uniformly distributed in the circle of radius one and centered at origin.

As the area of the disk is π, the joint density of V_1 and V_2 is:

$$f_{V_1 V_2}(v_1, v_2) = \frac{1}{\pi}.$$

Let R and Θ denote the polar coordinates of (V_1, V_2). As the Jacobian of the transformation is $\frac{1}{r}$, the joint density of R and Θ is given with:

$$f_{R\Theta}(r, \theta) = \frac{r}{\pi}. \tag{5.52}$$

Since the disk has area πr^2:

$$F_R(r) = \frac{\pi r^2}{\pi} = r^2, \tag{5.53}$$

and therefore:

$$f_R(r) = 2r. \tag{5.54}$$

Hence, R and Θ are independent with Θ uniformly distributed over $[0, 2\pi]$ and density function $f_\Theta = \frac{1}{2\pi}$. The sine and cosine of the random angle Θ can be generated by setting:

$$\begin{aligned}
\cos\Theta &= \frac{V_1}{R} = \frac{V_1}{\sqrt{V_1^2 + V_2^2}}, \\
\sin\Theta &= \frac{V_2}{R} = \frac{V_2}{\sqrt{V_1^2 + V_2^2}}.
\end{aligned} \tag{5.55}$$

Setting $S = R^2$, it is easy to show that S is uniformly distributed over $[0, 1]$ and independent of Θ. A pair X, Y of independent, normally distributed random variables can be generated as follows:

$$
\begin{aligned}
X &= \sqrt{-2 \log S} \, \frac{V_1}{\sqrt{S}} = V_1 \sqrt{\frac{-2 \log S}{S}}, \\
Y &= \sqrt{-2 \log S} \, \frac{V_2}{\sqrt{S}} = V_2 \sqrt{\frac{-2 \log S}{S}}.
\end{aligned}
\tag{5.56}
$$

A M-dimensional random vector $w = (w_1, w_2, \cdots w_M)^T$ is said to have a multidimensional normal distribution if, and only if, every non-trivial combination of the M components of w has a univariate normal distribution. The distribution of w is denoted by $\mathcal{N}(\mu, \Sigma)$, where μ is a $M \times 1$ mean vector with entries $\mu_m = E[w_m]$ and Σ is a $M \times M$ covariance matrix whose $(m, n)^{th}$ entry is $\mathrm{cov}[w_m, w_n]$.

The density function of w with support $\mathbb{R}^M$ is given with (see for example Johnson (1987), p.50):

$$
f(w) = \frac{1}{(2\pi)^{\frac{M}{2}} \, |\, \Sigma\, |^{\frac{1}{2}}} \exp\left[-\frac{1}{2}(w - \mu)^\top \Sigma^{-1}(w - \mu) \right],
\tag{5.57}
$$

where $|\, \Sigma\, |$ is the determinant of the non-singular matrix Σ.

The distribution of w can be represented as a linear transformation of M independent normally distributed variables $X = (X_1, X_2, \cdots, X_M)^\top$:

$$
w = AX + \mu,
\tag{5.58}
$$

where A denotes a $M \times M$ matrix for which $AA^\top = \Sigma$.

The generation of variables with distribution $\mathcal{N}(\mu, \Sigma)$ is then straightforward. X can be generated by M successive calls to an univariate normal generator and $\mathcal{N}(\mu, \Sigma)$ is achieved by the linear transformation $AX + \mu$. The matrix A is not unique and a frequent choice is the lower triangular matrix L with $LL^\top = \Sigma$ (Choleski decomposition).

The Student-t Distribution A Student-t distributed variable w with mean μ and covariance Σ can be obtained, in the one- and in the multi-dimensional case and for an integer number of degrees freedom ν, as follows (see for instance Johnson (1987), p.118):

$$
w = \frac{X}{\sqrt{\frac{S}{\nu}}} + \mu,
\tag{5.59}
$$

where X and S are independent, $X \sim \mathcal{N}\left(0, \frac{\nu-2}{\nu} \Sigma\right)$, and $S \sim \chi^2_{(\nu)}$, i.e. $S = Z_1^2 + Z_2^2 + \cdots + Z_\nu^2$ with the Z_i independent and $\sim \mathcal{N}(0, 1)$.

5.3 Monte Carlo Simulation

The Monte Carlo (MC) simulation is a standard procedure used to investigate distribution problems which are analytically intractable. The availability of increasing computer power has contributed to the widespread use of the method. Conceptually the Monte Carlo simulation, in its standard form called crude Monte Carlo, is straightforward and may be summarised as follows:

- generation of a set of joint scenarios $w^1, \cdots, w^N$ for the values of the risk factors at time T with a pseudo- or quasi-random algorithm as discussed in Sect. 5.3,
- full valuation of the portfolio for each generated scenario w^n, which yields a set of realisations $g(w^n)$ for the portfolio value $g[w(T)]$,
- estimation of the portfolio distribution F_g by weighting each scenario $g[w^n]$ with $\frac{1}{N}$.

Hence, the Monte Carlo method approximates F_g with a discrete distribution F_g^{MC} the support of which is a set of N realisations for the portfolio values. The precision of the Monte Carlo estimates is directly related to N as it will be discussed below. For the pseudo-random method, by assuming that the generated sequence is truly random, the number of dimensions M has no influence on the accuracy of the estimates. The numerical effort involved in the Monte Carlo Simulation depends on the level of accuracy that should be achieved and hence on N. The computation time needed to generate the scenarios is only a small part of the total effort which is dominated by the evaluation of the portfolio for each scenario. It should be pointed out that the value functions of numerous financial instruments, especially of derivatives, have no analytically closed-form solutions and should be approximated through numerical methods. The *error analysis* presented in Sect. 5.3.1 indicates the convergence behaviour of Monte Carlo estimates and therefore the number of scenarios which have to be generated and evaluated in order to obtain a prescribed level of accuracy. Section 5.3.2 discusses common methods used to *reduce the variance* of the Monte Carlo estimates with the objective of speeding up the convergence of the estimates.

A derivative method of the Monte Carlo Simulation called *Delta-Gamma Monte Carlo* consists in the replacement of the actual risk profile by a global second order Taylor approximation. In case the risk factors are jointly normal distributed, the Delta-Gamma Monte Carlo approximation of F_g converges to the Delta-Gamma approximation $F_g^{\Delta-\Gamma}$ as $N \to \infty$. Unlike the Delta-Gamma method, Delta-Gamma Monte Carlo method poses no assumptions on the distribution of the risk factors. The adequacy of the Delta-Gamma Monte Carlo method depends obviously on the accuracy of the global quadratic approximation of the risk profile.

5.3.1 Error Analysis

The Mean $E[g[w(T)]]$ The accuracy of both the pseudo- and quasi-random method is well documented for the estimation E_g^{MC} of the mean $E[g[w(T)]]$:

$$E_g^{MC} = \frac{1}{N} \sum_{n=1}^{N} g[w^n] \qquad (5.60)$$

For the pseudo-random method probabilistic arguments are used conjointly with the idealised assumption that the generated sequence $w^1 \cdots w^N$ is truly random, i.e. represents an IID sample from the risk factor distribution. From the properties of independence, the sequence $g[w^1] \cdots g[w^N]$ is also IID and represents a random sample from the portfolio distribution. Assuming that $E[g[w(T)]]$ exists, the *strong law of large numbers* asserts that (see appendix A.5):

$$E_g^{MC} \overset{a.s.}{\to} E[g[w(T)]], \qquad (5.61)$$

i.e.:

$$P\left(\lim_{N\to\infty} E_g^{MC} = E[g[w(T)]] \right) = 1 \qquad (5.62)$$

and therefore E_g^{MC} is a consistent estimate of $E[g[w(T)]]$. Assuming further that the variance of $g[w(T)]$ exists and is equal to σ_g^2 it is easy to verify that the expectation of the squared value of the error is:

$$
\begin{aligned}
E\left[\text{ error }\right]^2 &= E\left[E_g^{MC} - E[g[w(T)]] \right]^2 \\
&= E\left[\frac{1}{N} \sum_{n=1}^{N} \left(g[w^n] \right) - E[g[w(T)]] \right]^2 \\
&= \frac{1}{N^2} E\left[\sum_{n=1}^{N} \left(g[w^n] \right) - N E[g[w(T)]] \right]^2 \qquad (5.63) \\
&= \frac{1}{N^2} N \ \text{var} \ [g[w(T)]] \\
&\qquad \textit{since the } g[w^n] \textit{ are independent} \\
&= \frac{1}{N} \sigma_g^2,
\end{aligned}
$$

and therefore, on average, the magnitude of the error is given by:

$$E\left| E_g^{MC} - E[g[w(T)]] \right| = N^{-\frac{1}{2}} \sigma_g. \qquad (5.64)$$

The *central limit theorem* (see appendix A.5) provides further information about the error of the pseudo-random Monte Carlo estimate:

$$\sqrt{N}\left[E_g^{MC} - E[g[w(T)]]\right] \xrightarrow{d} \mathcal{N}(0, \sigma_g^2) \tag{5.65}$$

or equivalently:

$$\sum_{n=1}^{N} \frac{g[w^n] - E[g[w(T)]]}{\sqrt{N}\sigma_g} \xrightarrow{d} \mathcal{N}(0, 1) \tag{5.66}$$

so that, for large N, the $1 - \alpha$ confidence interval for $E[g[w(T)]]$ is given with:

$$\left[E_g^{MC} - z_{\frac{\alpha}{2}}\left(\frac{s_{E_g^{MC}}}{\sqrt{N}}\right), E_g^{MC} + z_{\frac{\alpha}{2}}\left(\frac{s_{E_g^{MC}}}{\sqrt{N}}\right)\right] \tag{5.67}$$

where: $s_{E_g^{MC}} = \sqrt{\frac{1}{N-1}\sum_{n=1}^{N}\left[g[w^n] - E_g^{MC}\right]^2}$ and $z_{\frac{\alpha}{2}}$ denotes the $1 - \frac{\alpha}{2}$ percentile of the standard normal density.

From (5.66), the error converges in law to the Gaussian distribution. For a fixed N, the Gaussian approximation of the error distribution may be arbitrarily poor. If the absolute third moment ρ with:

$$\rho = E\big[|\, g[w(T)] - E[g[w(T)]]\, |\big]^3$$

is also finite, then the *Berry-Esséen theorem* establishes that the convergence is uniform (see for instance Ferguson (1996), p.31):

$$|\, \check{F}_N(x) - \phi(x)\, | < \frac{c\rho}{\sqrt{N}\sigma_g^3}, \tag{5.68}$$

where $\check{F}_N$ is the distribution function of:

$$\sum_{n=1}^{N}\left(\frac{g[w^n] - E[g[w(T)]]}{\sqrt{N}\sigma_g}\right),$$

$\phi(x)$ denotes the distribution function of $\mathcal{N}(0, 1)$, and c is a constant greater than 0.4097 and smaller than 0.7975.

As discussed in (5.63), the expected absolute value of the error is on average $N^{-\frac{1}{2}}\sigma_g$ and therefore the pseudo-random Monte Carlo estimates converge on average at a rate of $\sqrt{N}$. Hence the number of scenarios required to yield a given error e is:

$$N = \frac{\sigma_g^2}{e}, \tag{5.69}$$

and depends on the variance of $g[w(T)]$. Note that the convergence of the error is slow ($\approx N^{-\frac{1}{2}}$) but does not depend on M.

For the quasi-random Monte Carlo, the *Koksma-Hlawka* inequality provides a deterministic upper error bound:

$$\left| E_g^{MC} - E[g[w(T)]] \right| \leq V(g)D_N^*(u^1, \cdots, u^N).\qquad(5.70)$$

$D_N^*(u^1, \cdots, u^N)$ is the star discrepancy and $V(g)$ denotes the variation in the sense of Hardy and Krause, i.e. a measure of smoothness of $g[w(T)]$ (for a discussion see Niederreiter (1992), pp.18-21). Inequality (5.70) requires $V(g)$ to be bounded.

Hence the low-discrepancy method yields a deterministic error bound whose order of magnitude is dependent of the number of dimensions M. From inequalities (5.32) and (5.70), it results that the convergence of the absolute error of the quasi-random Monte Carlo estimate is of order $N^{-1}(\log N)^M$. Table 5.3 compares the order of magnitude of the error of the (idealised) pseudo-random and the quasi-random Monte Carlo estimates for several sequence lengths N and dimension sizes M. According to table 5.3, the quasi-random

N	Pseudo MC	Quasi MC			
		$M{=}1$	$M{=}2$	$M{=}5$	$M{=}10$
10^3	0.032	0.003	0.009	0.243	59.0
10^4	0.01	0.0004	0.0016	0.1024	104.9
10^5	0.0032	0.00005	0.00025	0.03125	97.7
10^6	0.001	0.000006	0.000036	0.0078	60.5

Table 5.3. Monte Carlo: magnitude of the error

Monte Carlo scheme is only then superior to the pseudo-random Monte Carlo algorithm when M is small, provided that N is sufficiently large. For moderate M, the advantage is no longer practical. For instance, for $M = 10$:

$$\frac{(\log N)^M}{N} < \frac{1}{\sqrt{N}}$$

for $N \approx 10^{40}$. The results of table 5.3 are, however, misleading. Firstly, a (deterministic) pseudo-random sequence is not truly random and the order of magnitude of its error depends also on M (see Niederreiter (1992), pp.169-170 for a discussion). Secondly, and most importantly, the error bound for the quasi-random method, as given by the Koksma-Hlawka inequality, is very "loose" and too conservative for most functions g. A detailed discussion about the error analysis for the quasi-random Monte Carlo method can be found in Ökten (1998).

Figure 5.8 displays the quasi-(Sobol) and pseudo-random Monte Carlo estimates of $E[g[w(T)]]$ for the portfolio 23-dimensional PF^{II} (see Sect. 1.6) and shows that the quasi-random algorithm, as described in Press and Teukolsky (1989), is significantly superior to the pseudo-random scheme. The estimations of the mean of all test portfolios described in Sect. 1.6 yield similar results.

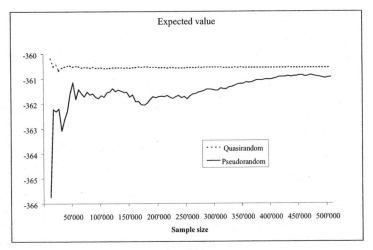

Figure 5.8. Pseudo- and quasi-random Monte Carlo: estimation of $E[g[w(T)]]$

Value-at-Risk Since the estimation of risk measures involves the whole portfolio distribution, the error analysis of $E[g[w(T)]]$ is clearly insufficient to assess the accuracy of a method. Figure 5.9 plots the pseudo- and quasi-random Monte Carlo estimates of VaR$_\alpha$, with $\alpha = 1\%$, 3% and 5%, for portfolio PF^{II}. Although the Sobol scheme remains superior to the pseudo-random algorithm, the advantage of the quasi-random Monte Carlo method is much less pronounced for the approximation of Value-at-Risk than for the estimation of the mean. The results are similar for all test portfolios. While probabilistic arguments may be used to assess the accuracy of pseudo-random estimates of VaR, very few results can be found in the literature about the quasi-random estimation of statistics other than the mean.

The derivation of confidence intervals for the pseudo-random estimates of Value-at-Risk, based on *order statistics*, will be presented in Sect. 6.2.3. An alternative approach, based on the *theorem of large deviations* allows the evaluation of the number of scenarios necessary for the estimation of VaR and will be presented in the sequel. According to Duffie and Pan (1997), let's assume that an event occurs with probability α (i.e. a loss equal or greater than VaR$_\alpha$) and that a random sample of size N is available to estimate α. What is the likelihood that the estimate $\hat{\alpha}_N$ of α will be greater than α' with $\alpha' > \alpha$? In other words, how high is the likelihood that the probability of

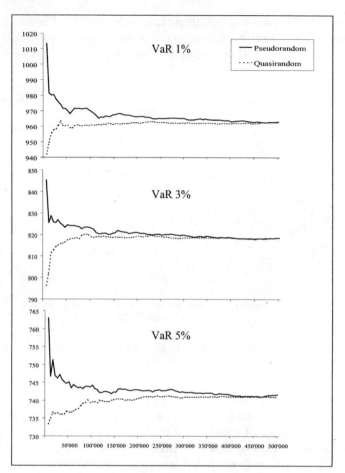

Figure 5.9. Pseudo- and quasi-random Monte Carlo: estimation of VaR

the event is overestimated ?

Let Y^n denote a Bernoulli random variable with an outcome of 1 if the nth evaluation $g[w^n]$ is greater than the α quantile and zero otherwise. The probability that Y^n equals one is α and $E(Y) = \alpha$. The estimate of α is obviously:

$$\hat{\alpha}_N = \frac{\sum_{n=1}^{N} Y_n}{N}. \tag{5.71}$$

The theorem of large deviations (see Durrett (1996), p.71, appendix A.5) states that:

$$P(\hat{\alpha}_N \geq \alpha') \leq e^{-N\gamma(\alpha')}, \tag{5.72}$$

where:

$$\gamma(\alpha') = \alpha'\theta - \log \Psi(\theta), \tag{5.73}$$

for an arbitrary $\theta > 0$. $\Psi(\theta) = E[e^{\theta Y^n}]$ is the moment generating function of Y^n.

The moment generating function $\Psi(\theta)$ of a Bernouilli variable Y is given with:

$$\sum_y e^{\theta y} p(y) = e^{0\theta}(1-\alpha) + e^{1\theta}(\alpha) = 1 - \alpha + e^\theta \alpha, \tag{5.74}$$

and (5.73) becomes:

$$\gamma(\alpha') = \alpha'\theta - \log \left[1 - \alpha + e^\theta \alpha\right]. \tag{5.75}$$

Since θ can be chosen freely, an optimal upper bound for $P(\hat{\alpha}_N \geq \alpha')$ can be obtained by minimising $e^{-N\gamma(\alpha')}$, i.e. maximising $\gamma(\alpha')$, with respect to θ and from (5.73) it results that:

$$\alpha' = \frac{\Psi'(\theta)}{\Psi(\theta)} = \frac{e^\theta \alpha}{1 - \alpha + e^\theta \alpha}, \tag{5.76}$$

which implies:

$$\theta = \log[\alpha'(1-\alpha)] - \log[\alpha(1-\alpha')],$$

and therefore 5.75 may be rewritten:

$$\gamma(\alpha') = \alpha' \left[\log \alpha'(1-\alpha)\right] - \log[\alpha(1-\alpha')] - \log \left[1 - \alpha + \frac{\alpha'(1-\alpha)}{1-\alpha'}\right]$$
$$= \alpha' \log \alpha' + (1-\alpha')\log(1-\alpha') - \alpha' \log \alpha + (\alpha'-1)\log(1-\alpha). \tag{5.77}$$

As an illustration, suppose that the statistic to estimate is the 95% percentile of F_g. The 1% likelihood that the estimate is greater than the true 96% percentile is approximated by:

$$1\% \approx e^{-N\gamma(\alpha')}$$

As $\gamma(\alpha') \approx 0.00113$ for $\alpha = 0.95$ and $\alpha' = 0.96$, the required number of scenarios is given with:

$$N \approx -\frac{\log(0.01)}{0.00113} \approx 4'087.$$

For $\alpha' = 0.97$, 943 scenarios are requested to achieve the 99% confidence interval, i.e. $P(\hat{\alpha}_N \geq \alpha') \leq 1\%$.

The Distribution Function $F_g(x)$ Again, probabilistic arguments are available to assess the accuracy of the pseudo-random Monte Carlo estimation of the distribution F_g. Assuming, as above mentioned, that the generated sequence is truly random, the set $g[w^1], \cdots, g[w^N]$ represents a random sample of F_g and the *sample or empirical distribution function* F_g^{MC} with:

$$F_g^{MC}(x) = \frac{1}{N} \sum_{n=1}^{N} I_{\{g[w^n] \le x\}} \tag{5.78}$$

is the Monte Carlo estimation of $F_g(x)$. The *Glivenko-Cantelli theorem* states that (see for instance Embrechts et al. (1997), p.62 or appendices A.4 and A.5):

$$P\left(\sup_{x \in \mathbb{R}} \mid F_g^{MC}(x) - F_g(x) \mid \to 0 \right) = 1 \tag{5.79}$$

Hence, the empirical distribution function converges uniformly and almost surely to the true distribution function and therefore $F_g^{MC}(x)$ is a consistent estimate of $F_g(x)$. Although 5.79 shows that the empirical distribution possesses the two enviable properties, i.e. consistency and uniform convergence, the Glivenko-Cantelli theorem provides few information about the number of scenarios necessary to achieve a prescribed level of accuracy.

Further insight is given by the *Kolmogorov-Smirnov statistic D_N*, which is defined as:

$$D_N = \sup_x \left| F_g^{MC}(x) - F_g(x) \right| = \max(D_N^+, D_N^-) \tag{5.80}$$

where:

$$\begin{aligned} D_N^+ &= \sup_x \{ F_g^{MC}(x) - F_g(x) \} \\ D_N^- &= \sup_x \{ F_g(x) - F_g^{MC}(x) \} \end{aligned} \tag{5.81}$$

Note that D_N represents the maximal distance between $F_g(x)$ and $F_g^{MC}(x)$ and is equivalent to the one-dimensional star discrepancy. It can be shown that for every $z > 0$ (see Gibbons (1971), p.81):

$$\lim_{N \to \infty} P\left(D_N \le \frac{z}{\sqrt{N}} \right) = L(z), \tag{5.82}$$

where:

$$L(z) = 1 - 2 \sum_{i=1}^{\infty} (-1)^{i-1} e^{-2i^2 z^2}. \tag{5.83}$$

The above statements apply for every continuous distribution. Equation (5.82) may be used as an approximation with sufficient accuracy for $N > 35$ and allows the determination of the minimum sample size N required to achieve a given level of accuracy ϵ, i.e. a sufficiently small D_N, for a given confidence level α. In other words, the minimal N, so that:

$$P(D_N \leq \epsilon) = \alpha. \tag{5.84}$$

Table 5.4 displays the values of α for several N and D_N. The right-hand side of (5.83) has been computed with the approximation given in Press et al. (1992), p.626.

Considering an (approximative) confidence level of 95%, the minimal sample

N	$\epsilon = 0.05$	$\epsilon = 0.03$	$\epsilon = 0.01$
10	4.74E-10	1.42E-09	3.67E-09
1000	0.99	0.67	3.48E-05
10000	≈ 1	0.99	0.73

Table 5.4. Kolmogorov-Smirnov statistic: confidence levels α for different N

size N is 750 for $D_N = 0.05$, 2'000 for $D_N = 0.03$ and 18'500 for $D_N = 0.01$. Once again, accuracy dictates large sample sizes. Moreover, the Kolmogorov-Smirnov statistic assumes that $g[w^1] \cdots g[w^N]$ represents a truly random sample of $F_g(x)$. To asses the accuracy of both the pseudo- and the quasi-random Monte Carlo methods, a portfolio with linear pay-off profile $g = (1, \cdots, 1)$ and jointly normal distributed risk factors $\mathcal{N}(0, \frac{I}{M})$, where I is the $M \times M$ identity matrix, will be constructed for $M = 5, 10, 20$ and 50. From (5.8) it is easy to verify that for every M, the profit-and-loss distribution yields the standard normal distribution. Figures 5.10 and 5.11 depict the pseudo-random, respectively quasi-random, approximations of $F_{\bar{g}}$.

Table 5.5 displays the empirical D_N of both methods for several M and $N = 10'000$. According to the numbers in table 5.4, D_N should be "almost surely" smaller than 0.05 and smaller than 0.01 with a probability of about 73%. The results show that all empirical D_N are smaller than 0.05 and that 5 out of 8 estimated D_N, i.e. about 63%, are smaller than 0.01. The results indicate further that the quasi-random scheme is superior to its pseudo-random equivalent for almost all M. Note that for a truly random sequence the size of M has no influence on the size of D_N. However, table 5.5 shows that both methods, but particulary the quasi-random scheme, deteriorate with increasing M.

5.3.2 Variance Reduction Techniques

As seen in the last section, the number of scenarios to be generated in order to achieve an adequate level of accuracy can lie in the thousands. To reduce the

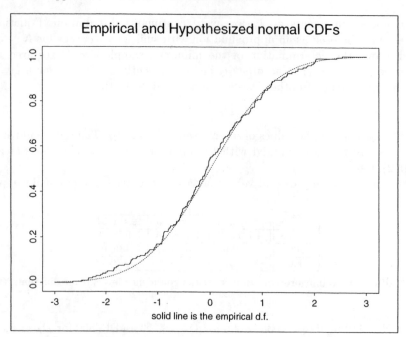

Figure 5.10. Linear portfolio: pseudo-random Monte Carlo

M	5	10	20	50
Pseudo-random Monte Carlo	0.0075	0.0080	0.0076	0.0133
Quasi-random Monte Carlo	0.0053	0.0071	0.0084	0.0110

Table 5.5. Kolmogorov-Smirnov statistic of MC estimates ($N = 10'000$)

computational burden of the Monte-Carlo method, several variance reduction approaches have been proposed and are widely used for the estimation of $E[g[w(T)]]$. In Sect. 5.3.1 it has been shown that the estimation error of the expected value is, on average, $\sigma_g/\sqrt{N}$ for the pseudo-random Monte Carlo method. A method which allows the decrease in volatility of the estimator by a factor 10 will be equivalent in terms of error reduction to an increase in the number of samples by a factor of 100.

The variance reduction methods for the pseudo-random scheme can be grouped into three broad classes (see for instance Kalos and Whitlock (1986), pp.89-116 and Berman (1998)):

- *Correlation methods*
  The correlation methods include both the *control variable* and the *antithetic variable* techniques. The control variable technique supposes the existence of a similar function $g'[w(T)]$ with known expected value to con-

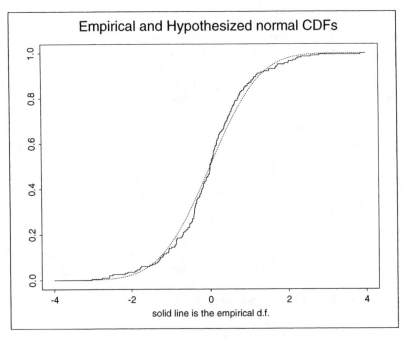

Figure 5.11. Linear portfolio: quasi-random Monte Carlo

struct a new unbiased estimator of $E\big[g[w(T)]\big]$. The antithetic variable technique generates pairs $[w^n, w_A^n]$ of scenarios which are negatively correlated with the goal that the pairs $[g[w^n], g[w_A^n]]$ are in turn negatively correlated and therefore reduce the variance of the Monte Carlo estimates (see below).

- *Importance sampling*
  The methods generate large scenario samples in regions where the function $g[w(T)]$ is large or rapidly varying and smaller samples in the regions where $g[w(T)]$ exhibits little variation.

- *Stratified sampling*
  Theses methods partition the risk factor space into groups (strata) and generate separately scenarios for each stratum. The number of generated scenarios assigned to each group may vary. An estimator is defined for each stratum and the overall estimator is a weighted average of the single estimators. The weights depend on the probability of each stratum. This method reduces the variance of the estimator significantly when, with respect to the value function, the strata are relatively homogenous but there are considerable differences between the strata.

Variance reduction methods which do not alter the order of low-discrepancy sequences can be used in conjunction with the quasi-random methods. As discussed in Galanti and Jung (1997), Joy et al. (1996) and Fox (1986), importance sampling and control variables may improve the quasi-random scheme. Antithetic variables change the order of the sequences and do not improve the basic quasi-random Monte Carlo scheme.

Variance reduction supposes known information about the problem as it relies on specific characteristics of the value function $g[w(T)]$ over the whole probability space. In the context of risk management the value function changes steadily with the permanent shifts in the portfolio structure. The possible reduction in the computational effort achieved with the variance reduction methods should therefore be compared with the additional analytical work required to analyse the problem for each change in the portfolio structure.

Only the antithetic variable technique may be implemented in a way, which does not necessitate any information about $g[w(T)]$. Supposing, pairs such as $(w^n, w^n_A; n = 1, 2, \cdots, N)$ with identical distribution can be generated so that each pair is negatively correlated, both estimators:

$$\frac{1}{N} \sum_{n=1}^{N} g[w^n], \qquad \frac{1}{N} \sum_{n=1}^{N} g[w^n_A] \qquad (5.85)$$

are unbiased estimators of the expected value. It results that following statistic:

$$\frac{1}{N} \sum_{n=1}^{N} \frac{g[w^n] + g[w^n_A]}{2} \qquad (5.86)$$

is an unbiased estimator of the mean as well. Its variance is given with:

$$\begin{aligned}
\mathrm{var}\left[\frac{g[w^n] + g[w^n_A]}{2}\right] &= \frac{1}{4} \, \mathrm{var} \, g[w^n] + \frac{1}{4} \, \mathrm{var} \, g[w^n] + \frac{1}{2} \, \mathrm{cov}\left[g[w^n], g[w^n_A]\right] \\
&= \frac{\mathrm{var}\left[g[w^n]\right] + \mathrm{cov}\left[g[w^n] + g[w^n_A]\right]}{2}
\end{aligned} \qquad (5.87)$$

since $g[w^n]$ and $g[w^n_A]$ have same variance.

An obvious choice of w^n and w^n_A for risk factor distributions symmetric about the mean μ is, given $w^n = \mu + x$:

$$w^n_A = \mu - x \qquad (5.88)$$

where $w^n = w^n_1, \cdots w^n_M$, $\mu = \mu_1, \cdots, \mu_M$ and $x = x_1 \cdots x_M$. Hence (w^n, w^n_A) represents a pair of scenarios with risk factor distribution and correlation matrix $-I$, where I is the identity matrix. If $g[w^n]$ and $g[w_A^n]$ are strongly negatively correlated, the method is effective in reducing the variance of the estimate. However, it should be noted that a negative correlation of w^n and w_A^n does not imply a negative correlation of $g[w^n]$ and $g[w_A^n]$. If for instance $g[w(T)]$ is symmetric about μ no variance reduction will be achieved.

5.4 The BDPQA

Scenario-based methods are versatile enough to adequately model the risk factor dynamics. The results of Sect. 5.3 suggest, however, that accurate Monte Carlo estimates of the portfolio distribution or of risk measures derived from F_g require thousands of scenarios. As discussed above the associated numerical effort lies mostly in the number of full evaluations which have to be computed. This section presents a new approach which allows to capture the information of large sets of scenarios for the risk factors and to mitigate the computational burden associated with scenario-based methods. The approach, called *Barycentric Discretisation with Piecewise Quadratic Approximation* or BDPQA for short, replaces the actual risk profile by a piecewise quadratic approximation, as discussed in Sect. 4.5.2. The calculation of the BDPQA involves following steps:

- generation of a set S_1 of joint scenarios $w^1, \cdots, w^{N^1}$ for the values of the risk factors at time T with a pseudo- or quasi-random algorithm as for the Monte Carlo method,

- generation of a set S_2 of distinguished joint scenarios $\Lambda^1, \cdots, \Lambda^J$, called *generalised barycenters*,

- approximation of the risk profile $g^\nabla[w^n]$ for each scenario w^n by the means of piecewise quadratic approximation whose supporting point is an element Λ^j of S_2,

- estimation F_g^∇ of the portfolio distribution F_g by weighting each scenario $g^\nabla[w^n]$ with $\frac{1}{N^1}$

Hence the BDPQA differs principally from the Monte Carlo method through the generation of a second set of scenarios S_2 which serve as supporting points for the piecewise quadratic approximation of the actual risk profile. The choice of the elements of S_2 is based on the *barycentric discretisation* which has been developed in Frauendorfer (1992, 1996) in the more general context of the barycentric approximation methodology used in the field of stochastic programming. The barycentric approximation helps to overcome the problem of multidimensional integration by discretisation of the stochastic processes. Based on extensions of Jensen's and Edmundson-Madansky inequalities, the Barycentric approximation provides a lower and upper bound in case of convex or concave g.

5.4.1 Simplices

In this section, which follows in part the conceptual layout of Bazaraa et al. (1993), ch.2, basic concepts of euclidian geometry are recalled and the properties of simplices are considered.

A *neighbourhood* of a point $x \in R^M$ (or the ball with radius ε and center x) is the set

$$\{y \in R^M : \| y - x \| < \varepsilon\}, \tag{5.89}$$

where ε is some positive number.

A point $x \in S$ is said to be an *interior point* if all points within some neighbourhood of x are also in S. A point x is said to be a *boundary point* of S if every neighbourhood of x contains a point in S and not in S. The set of all boundary points of S is the *boundary* of S denoted by ∂S, the set of all interior points of S is called the *interior* of S and denoted by $\text{int}(S)$. The set S is said to be open if it contains no boundary points and *closed* if it contains its boundary. A set that is contained in a ball of finite radius is called a bounded set. A bounded and closed set is compact.

A set $S \in R^M$ is said to be *convex* if the line segment joining any two points of the set belongs to the set. Weighted averages of the form $\sum_{j=1}^{k} \lambda_j x_j$ where $\sum_{j=1}^{k} \lambda_j = 1, \lambda_j \geq 0$ and the $x_j \in S$ are referred to as *convex combinations* of the x_j. If the non-negativity conditions on the multipliers λ_j are dropped, the combinations are called *affine combinations*.

The *convex hull* of an arbitrary set $S \in R^M$, denoted by $H(S)$ or $\text{conv}(S)$ is the collection of all convex combinations of S. In other words, $x \in H(S)$ if, and only if, x can be represented as:

$$x = \sum_{j=1}^{k} \lambda_j x_j,$$

where:

$$\sum_{j=1}^{k} \lambda_j = 1 \tag{5.90}$$

$$\lambda_i \geq 0 \text{ for } j = 1, \cdots, k.$$

$H(S)$ is the smallest convex set containing S, it is indeed the intersection of all convex sets containing S.

A *hyperplane* $H \in R^M$ is a collection of points of the form $\{x : p^\top x = \alpha\}$ where p is a non-zero vector in R^M and α a scalar. The hyperplane H divides R^M into two *closed halfspaces* denoted by H_+ and by H_- with:

$$H_+ = \{x : p^\top x \geq \alpha\}$$
$$H_- = \{x : p^\top x \leq \alpha\}$$

A hyperplane H is called a *supporting hyperplane* of S at a boundary point x' of S if either $S \subseteq H_+$, that is $x : p^\top(x - x') \geq 0$, or $S \subseteq H^-$, that is $\{x : p^\top(x - x') \leq 0\}$. A *convex polyhedron* CP is the intersection of a finite number of closed halfspaces in R^M. CP is a D-dimensional polyhedron (D-polyhedron) if the points in CP span R^D, i.e. if any vector in R^D can be represented as a linear combination of the points in CP. A face F of a D-polyhedron CP is the intersection of CP with a supporting hyperplane. F itself is a polyhedron of some lower dimension. If the dimension of F is K we call F a K-face of CP. 0-faces of CP are called *vertices*, 1-faces are called *edges*. The 0-faces of a convex set S are also called the extreme points of the set. If x is an extreme point of the set S:

$$x = \lambda x_1 + (1 - \lambda)x_2, \tag{5.91}$$

where:

$$x_1, x_2 \in S \text{ and } 0 < \lambda < 1,$$

it implies that:

$$x = x_1 = x_2. \tag{5.92}$$

Therefore, extreme points can not be expressed as convex combination of other points of the set.

A *convex D-dimensional polytope* is a bounded convex D-polyhedron. Alternatively a polytope is the convex hull of a finite number of points $x_1, \cdots, x_{K+1} \in R^M$. If $x_1, \cdots x_{K+1}$ are affinely independent, i.e. $x_2 - x_1, x_3 - x_1, \cdots, x_{K+1} - x_1$ are linearly independent, then $H(x_1, \cdots, x_{K+1})$ is called a *simplex* with vertices $x_1, \cdots, x_{K+1}$. As the maximum number of linearly independent vectors in R^M is M, there could be no simplex in R^M with more than $M+1$ vertices. As $x_1, \cdots x_{K+1}$ are affinely independent, K is the dimension of the simplex. A simplex of dimension M in R^M is called regular. By definition a point in the convex hull of a set can be represented as a convex combination of a finite number of points in the set. Since a simplex is a compact convex set, any point in the set can be represented as a convex combination of the extreme points or vertices of the simplex. The unique representation of a point x in a simplex S in relation to the vertices $v_1, \cdots v_{K+1}$ of the simplex are given with:

$$
\sum_{k=1}^{K+1} \lambda_k(x)v_k = x
$$

$$
\sum_{k=1}^{K+1} \lambda_k = 1,
$$

(5.93)

where $\lambda_k \geq 0$ for $k = 1, \cdots, K+1$.

The scalars $\lambda_k(x)$ are called *barycentric coordinates (weights)* of x with respect to the simplex. The point $x \in S$ with equal barycentric weights $\lambda_1 = \lambda_2, \cdots, = \lambda_{K+1}$ is called the *barycenter* of the simplex.

5.4.2 Simplicial Coverage of the Risk Factor Distribution

The BDPQA is based on a discretisation of the probability space of the risk factors through *simplicial coverage*. The method can be applied to any arbitrary distribution. For a probability distribution with finite supports, it will be possible to find a simplex covering all values of this distribution. For a probability distribution with infinite support, like the normal or Student-t distributions, only a percentage of all values can be covered by a simplex with finite vertices.

It is therefore necessary to approximate the probability space $(\Omega, \mathcal{B}, P)$ by $(\hat{\Omega}, \hat{\mathcal{B}}, \hat{P})$, where $\hat{\Omega}$ represents a simplex, $\hat{\mathcal{B}} := \{\hat{B} | \hat{B} = B \cap \hat{\Omega}, \forall B \in \mathcal{B}\}$, and:

$$
\hat{P}(B) = \frac{P(B \cap \hat{\Omega})}{P(\hat{\Omega})}.
$$

(5.94)

For a given sufficient small positive ϵ, a good simplicial coverage requires finding the smallest simplex $\hat{\Omega}$ for which $P(\hat{\Omega}) \geq 1 - \epsilon$. The simplicial coverage of the normal distribution as described in (Frauendorfer and Härtel (1996)) and extended to the Student-t distribution is presented in the sequel. Note that the method can be applied to all elliptically symmetric distributions (see Sects. 2.1.2 and A.3.2).

The Uncorrelated Case Let's consider the case of the M-dimensional uncorrelated standard normal and Student-t distributions, i.e. with mean 0 and a correlation matrix that equals the identity matrix. For both distributions, the confidence region is a closed ball centered at the origin with radius r. The coverage probability $1 - \epsilon$ determines the value of the radius. For the normal distribution $r = \Phi\left(1 - \frac{1-\epsilon}{2}\right)$, where $\Phi()$ denotes the cumulative distribution function of the standard normal distribution. For a coverage probability of 95%, r is therefore approximately 1.96. For the Student-t distribution, the value of r is a function of the number of degrees of freedom. The corresponding value of r for a Student-t distribution with four degrees of freedom is approximately 2.57. The coverage probability has to be determined heuristically. In the context of risk management, where typically extreme events with small probability are very important, a high coverage probability should be ensured. Empirically, the value of 4 for r in the normal case, which corresponds to a coverage probability superior to 99.99%, has provided good results.

In the one-dimensional case the simplicial coverage is the simplex spanned by the vertices $[-r, +r]$. Due to the symmetry around the origin of the confidence regions of both distributions, the multidimensional simplicial coverage corresponds to a simplex whose barycenter is located at the origin and whose edges are all of the same side length. The side length of the M-dimensional simplex can be calculated as:

$$c_M = \rho_M \cdot c_{M-1}, \tag{5.95}$$

where:

$$\rho_M = \frac{M+1}{\sqrt{\beta_{M-1}^2 - (M-1)^2}} \quad \text{and} \quad \beta_M = \beta_{M-1} \cdot \rho_M. \tag{5.96}$$

The proof by induction is given in Frauendorfer and Härtel (1996). A M-dimensional simplex can be constructed from the $M-1$ dimensional simplex using the side length c_M and setting the Mth coordinate equal to $\frac{1}{\beta_M c_M}$. The additional vertex is positioned with coordinates:

$$\left(0, \cdots, 0, -\frac{M}{\beta_M} \cdot c_M\right)^\top,$$

and hence the vertices of this simplex have the following form:

$$
\begin{pmatrix} \frac{1}{\beta_1} \cdot c_M \\ \frac{1}{\beta_2} \cdot c_M \\ \frac{1}{\beta_3} \cdot c_M \\ \vdots \\ \frac{1}{\beta_M} \cdot c_M \end{pmatrix},
\begin{pmatrix} -\frac{1}{\beta_1} \cdot c_M \\ \frac{1}{\beta_2} \cdot c_M \\ \frac{1}{\beta_3} \cdot c_M \\ \vdots \\ \frac{1}{\beta_M} \cdot c_M \end{pmatrix},
\begin{pmatrix} 0 \\ -\frac{2}{\beta_2} \cdot c_M \\ \frac{1}{\beta_3} \cdot c_M \\ \vdots \\ \frac{1}{\beta_M} \cdot c_M \end{pmatrix}, \cdots,
\begin{pmatrix} 0 \\ 0 \\ 0 \\ \vdots \\ -\frac{M}{\beta_M} \cdot c_M \end{pmatrix}
$$

Clearly, the smallest simplex which surrounds the confidence region is not unique. An infinite number of coverages can be found by rotating the simplex with respect to the coordinate axes. Figure 5.12 shows some examples of the two-dimensional case.

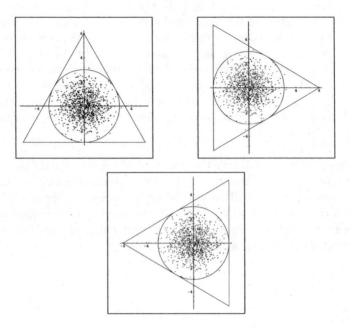

Figure 5.12. Simplicial coverage for the uncorrelated case

The Correlated Case As noted above, the distribution of a variable Y with distribution $\mathcal{N}_M(\mu, \Sigma)$ can be represented as a linear transformation of M independent normal deviates in $X = (X_1, X_2, \cdots X_M)^\top$:

$$Y = AX + \mu, \qquad (5.97)$$

where A any $M \times M$ matrix for which $AA^\top = \Sigma$.

Let L be the lower triangular matrix of the Choleski decomposition of Σ, i.e. $\Sigma = LL^\top$. The vertices of the simplicial coverage for Y are then:

$$v_m^Y = \mu + Lv_m^X \quad m = 1, \cdots, M + 1, \qquad (5.98)$$

where v_m^X is the mth vertex of the simplex covering the M-dimensional standard normal distribution. The same transformation applies for the Student-t case.

Obviously, the simplicial coverage in the correlated case depends on the choice of the simplex covering the uncorrelated distribution (see Fig. 5.13). It has been empirically verified that the barycentric discretisation is insensitive to the choice of the simplex surrounding the confidence region.

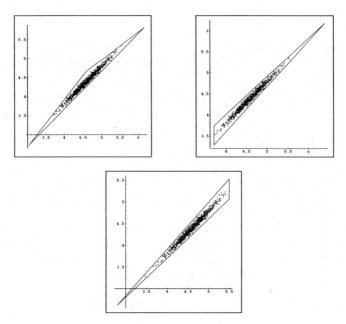

Figure 5.13. Simplicial coverage for the uncorrelated case

5.4.3 Barycentric Discretisation

Given a probability space $(\hat{\Omega}, \hat{\mathcal{B}}, \hat{P})$ with $\hat{\Omega}$ being a simplex, let Θ denote the set of those probability measures Q on $(\hat{\Omega}, \mathcal{B})$ which coincide in the first moments with those of $\hat{P}$; i.e. Θ consists of those probability measures Q for which:

$$\int_{\hat{\Omega}} w(T)dQ = \int_{\hat{\Omega}} w(T)d\hat{P}, \qquad (5.99)$$

holds.

A partial ordering (see Stoyan (1983)) $\leq^{(c)}$ for the set Θ may be defined with respect to the set $\mathcal{C}_{\hat{\Omega}}$ of continuous convex functions relative to the simplex $\hat{\Omega}$:

$$Q_1 \leq^{(c)} Q_2 \Leftrightarrow \int_{\hat{\Omega}} g[w(T)]dQ_1 \leq \int_{\hat{\Omega}} g[w(T)]dQ_2, \quad \forall g(\cdot) \in \mathcal{C}_{\hat{\Omega}}. \qquad (5.100)$$

The set of extremal probability measures of Θ taken with respect to $\leq^{(c)}$ is then defined according to:

$$\inf{}^{(c)}\Theta := \{Q_{inn}|Q_{inn} \leq^{(c)} Q, \quad \forall Q, Q_{inn} \in \Theta\}, \qquad (5.101)$$
$$\sup{}^{(c)}\Theta := \{Q_{out}|Q \leq^{(c)} Q_{out}, \quad \forall Q, Q_{out} \in \Theta\}, \qquad (5.102)$$

It is proven that the sets $\inf{}^{(c)}\Theta$, respectively $\sup{}^{(c)}\Theta$, are singletons (see Frauendorfer (1992)), are called *inner*, respectively *outer*, discretisation and denoted by $\hat{Q}_{inn}$, respectively $\hat{Q}_{out}$. The support of $\hat{Q}_{inn}$ is a singleton whose element is also called the *generalised barycenter* Λ of the simplex $\hat{\Omega}$. It is completely determined by the first moments which characterize the set Θ:

$$\Lambda = \int_{\hat{\Omega}} w(T)d\hat{P} \qquad (5.103)$$

where $w(T) = (w_1(T), \cdots, w_M(T))$. Clearly, the probability assigned to the generalised barycenter is one.

The support of $\hat{Q}_{out}$ is finite and consists of the $M+1$ vertices of the simplex $\hat{\Omega}$. The probabilities that are assigned to these vertices are the barycentric weights $\lambda_m (m = 1, 2, \cdots, M + 1)$ of the generalised barycenter and, hence, are completely determined by the first moments.

The barycentric weights of the generalised barycenter Λ are the solutions of the following linear equations:

$$\sum_{m=1}^{M+1} \lambda_m(\Lambda) = 1,$$
$$\sum_{m=1}^{M+1} v_m \lambda_m(\Lambda) = \Lambda. \qquad (5.104)$$

Let $\lambda(\Lambda)$ denote the $M + 1$ dimensional vector of the barycentric weights of Λ and V following $(M + 1) \times (M + 1)$ dimensional matrix:

$$\begin{pmatrix} 1 & 1 & \cdots & 1 \\ v_1 & v_2 & \cdots & v_{M+1} \end{pmatrix}$$

Equations (5.104) may be rewritten in matrix notation:

$$V\lambda(\Lambda) = \begin{pmatrix} 1 \\ \Lambda \end{pmatrix}, \qquad (5.105)$$

and therefore the unique solution:

$$\lambda(\Lambda) = V^{-1} \begin{pmatrix} 1 \\ \Lambda \end{pmatrix}. \qquad (5.106)$$

In order to achieve a finer subdivision of the probability space, both discretisations can be refined through the partition of $\hat{\Omega}$. An adequate point on a selected edge has to be chosen, subject to which the simplex is split in two subcells. The partition insures that both subcells are themselves simplices (usually called *subsimplices*) with mutually disjoint interior. After this first refinement, the support of $\hat{Q}_{inn}$ is given by the pair of generalised barycenters of both subsimplices. The initial generalised barycenter of the simplex is no longer an element of $\hat{Q}_{inn}$. However, the refinement does not affect the boundary of the simplex and the initial vertices still remain elements of the support of $\hat{Q}_{out}$. The point where the simplex has been split is a vertex of both subsimplices and is therefore an additional element of the support of $\hat{Q}_{out}$. For further refinements the choice of the subsimplex to be split must additionally be met. The refinement issue is discussed in Sect. 5.4.5.

Let $\mathcal{P}^J$ be a simplicial partition of $\hat{\Omega}$ achieved after $J-1$ refinements; i.e. $\mathcal{P}^J := \{\hat{\Omega}^j;\ j = 1, \cdots, J\}$ where the J subsimplices have mutually disjoint interiors and whose union equals $\hat{\Omega}$. Obviously:

$$\sum_{j=1}^{J} \hat{P}\left(\hat{\Omega}^j\right) = 1, \tag{5.107}$$

where:

$$\hat{P}\left(\hat{\Omega}^j\right) = \int_{\hat{\Omega}^j} w(T)d\hat{P}. \tag{5.108}$$

Applying the above statements to each of these subcells $\hat{\Omega}^j$ ($j = 1, \cdots, J$) the respective discretisations are noted with $\hat{Q}_{inn}^J$ and $\hat{Q}_{out}^J$.

The support of $\hat{Q}_{inn}^J$ are the J generalised barycenters Λ^j with respective probability weighting $\int_{\hat{\Omega}_j} w(T)d\hat{P}$ (see Fig. 5.14).

The support of $\hat{Q}_{out}^J$ are the $M+1+J$ vertices (see Fig. 5.15). The probability assigned to the vertex v_m^j of the subsimplex Ω^j is:

$$\hat{P}\left(\hat{\Omega}^j\right) \lambda_m^j. \tag{5.109}$$

Note that the number of generalised barycenters and vertices increases linearly with the number of partitions. If the diameters of all subcells tend towards 0 for $J \to \infty$ the weak convergence of $\hat{Q}_{inn}^J$ and $\hat{Q}_{out}^J$ to $\hat{P}$ is ensured.

5.4.4 Approximation of the Portfolio Distribution

Once the risk factor space has been discretised, the approximation of the portfolio distribution requires the evaluation or the approximation of the

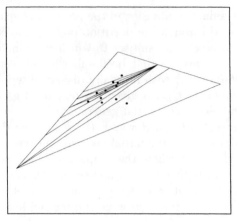

Figure 5.14. Support of $\hat{Q}_{inn}^{J}$ for $J = 10$

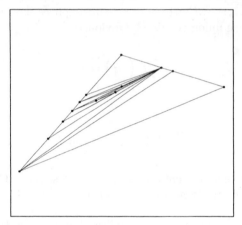

Figure 5.15. Support of $\hat{Q}_{out}^{J}$ for $J = 10$

price function for some values of $w(T)$. A natural choice is obviously the evaluation at the supports of both discretisations. The associate approximate distributions are given with:

$$\hat{F}_{g,inn}^{J}(x) = \hat{Q}_{inn}^{J}\left(g[w(T)] \leq x\right), \tag{5.110}$$

respectively:

$$\hat{F}_{g,out}^{J}(x) = \hat{Q}_{inn}^{J}\left(g[w(T)] \leq x\right). \tag{5.111}$$

$\hat{F}_{g,inn}^{J}$, the *inner approximation*, is given by the evaluation of g at the generalised barycenters.

The *outer approximation* $\hat{F}_{g,out}^{J}$ is achieved by the evaluation of g at the vertices. If the diameters of all subcells tend towards 0 for $J \to \infty$ it holds for each $x \in \mathbb{R}$ at which $F_g(x)$ is continuous:

$$\lim_{J \to \infty} \hat{F}^J_{g,inn}(x) = \lim_{J \to \infty} \hat{F}^J_{g,out}(x) = F_g(x). \tag{5.112}$$

However, empirical results related to the approximation of Value-at-Risk figures of portfolios with highly non-linear pay-off functions (see Frauendorfer and Königsperger (1995)) have shown that the convergence of $\hat{Q}^J_{inn}$ and $\hat{Q}^J_{out}$ is not strong enough to ensure an adequate convergence of the quantiles of $\hat{F}^J_{g,inn}$ and $\hat{F}^J_{g,out}$.

These results have called for the use of additional information as described in Frauendorfer et al. (1997). Instead of evaluating the portfolio at each supporting point of the discretisations, a linear approximation with respect to each subsimplex, i.e. a piecewise linear approximation, is computed. The results presented in Frauendorfer et al. (1997) show that the piecewise linear approximation and the Delta-Gamma Monte Carlo, which is based on a global quadratic approximation of the risk profile, are of similar accuracy. The better results, obtained by using the first order information, has motivated the author to consider higher order terms in order to capture the non-linearity of the risk profiles. As discussed in Sect. 4.5, the use of high order terms does not necessarily ensure the convergence of the Taylor approximation to the actual risk profile. Moreover, the numerical effort has to remain tractable and the information in form of order derivatives must be available. In practice, only the first and second order sensitivities of $g[w(T)]$ are at disposal. As a result, the BDPQA considers the second order approximation of the risk profile:

$$
\begin{aligned}
g^{\nabla,j}[w^n] = g[\Lambda^j] &+ \sum_{m=1}^{M} \frac{\partial g[\Lambda^j]}{\partial w_m}(w_m^{n,j} - \Lambda_m^j) \\
&+ \frac{1}{2!} \sum_{m_1=1}^{M} \sum_{m_2=1}^{M} \frac{\partial^2 g[\Lambda^j]}{\partial w_{m_1} \partial w_{m_2}}(w_{m_1}^{n,j} - \Lambda_{m_1}^j)(w_{m_2}^{n,j} - \Lambda_{m_2}^j),
\end{aligned}
\tag{5.113}
$$

for the N_j scenarios $w^{n,j}$ located in j^{th} subsimplex, the generalised barycenter of which is Λ^j. Note that:

$$\sum_{j=1}^{J} N_j = N^1.$$

The approximation $F_g^{\nabla,J}$ of F_g is therefore given with:

$$F_g^{\nabla,J}(x) = \frac{1}{N^1} \sum_{j=1}^{J} \sum_{n=1}^{N_J} I_{\{g^{\nabla,j}[w^n] \leq x\}}, \tag{5.114}$$

where I denotes the indicator function.

It should be noted that the approximation requires additional information of order 2 in the dimension M. For each supporting point the $M \times M$-dimensional Hessian matrix of g is needed in addition to the M-dimensional Gradient vector. As the Hessian is symmetric, the evaluation of $\frac{M(M+1)}{2}$ items is requested. The critical points are the availability of the Hessian matrix and of the Gradient vector for each $\varLambda^j$ as well as the number J of refinements required to obtain accurate estimates of F_g. If the information is available and J is moderate, i.e. $J << N^1$, the numerical effort associated with the BDPQA remains considerably less important than the N^1 full valuations of the Monte Carlo method.

5.4.5 Refinement Strategies

As stated above, within the context of the barycentric discretisation, the accuracy of the approximation is improved by subdividing the probability space through a simplicial partition of $\hat{\varOmega}$. The considered simplicial partition $\mathcal{P}^J$ is obtained by a sequence of successive refinements of $\hat{\varOmega}$. Given $\mathcal{P}^{J-1}$, $\mathcal{P}^J$ is achieved by splitting one subcell of $\mathcal{P}^{J-1}$. This requires the choice of an adequate subcell (from the second refinement up) of $\mathcal{P}^J$, of an edge of the subcell and of the point on the edge at which the split will be made.

The goal of the refinement is obviously the decrease in error due to the approximation of the risk profile. The numerical effort associated with it must remain affordable and depends clearly on the number of evaluations of g which have to be made. In the case of the BDPQA each refinement requires at least one additional evaluation of g. An adequate sequence of partitions may speed up the decrease in error and therefore the decrease in the number of refinements needed.

Within the context of stochastic linear programming, sophisticated refinement schemes have been developed in Frauendorfer and Marohn (1996) and Marohn (1998). However, the structural properties assumed there, concavity or convexity of the value functions, do not generally hold for risk profiles of portfolios containing options. Alternative simplicial partitions for risk management have been considered in Frauendorfer and Königsperger (1995). The strategies proposed there focus on the estimation of Value-at-Risk figures and therefore on specific regions of the portfolio distribution.

In the sequel, no structural properties of the value function g other than differentiability are assumed. Moreover, it is not focussed on a specific region of the portfolio distribution. In the case of the BDPQA, the error arises through the replacement of the exact value of g by a local quadratic approximation, i.e. by truncation of the Taylor approximation after the second order terms. For a given realisation w^n of $w(T)$ the accuracy of the approximation depends on the degree of "non-quadraticity", i.e. of the non-linearity of higher order than two, of g and on the location of the realisation of $w(T)$. For linear and quadratic functions, the approximation is exact. For functions with third and/or higher order terms, the approximation worsens with increasing distance between the realisations of $w(T)$ and the corresponding supporting points of the approximation. As the number of refinements is limited, the probability information should also be considered: it is more useful to achieve a good approximation in regions of the risk factor space with high probability weights, i.e. in regions where realisations of $w(T)$ are more likely. To sum up, for the construction of an adequate refinement strategy following information is of interest:

- the *probability information*,
  i.e. the probability weights of the subspaces of Ω,
- the *value information*,
  i.e the degree of non-linearity of higher order than two of g at different points in Ω and
- the *geometrical information*,
  i.e. the distance between the realisations w^j and the supporting points Λ^j of the approximation for a given partition $\mathcal{P}^J$ of Ω.

Within the framework of the barycentric discretisation the requested information is provided as follows:

- *Probability information*
  For a given partition of $\hat{\Omega}$, the probability weight of each subsimplex $\hat{\Omega}^j$, $j = 1, 2, \cdots J$ is given with:

$$\hat{P}\left(\hat{\Omega}^j\right) = \int_{\hat{\Omega}^j} w(T)d\hat{P}. \tag{5.115}$$

- *Value information*
  For a given partition $\mathcal{P}^j$ with $j \leq J$ the value function g and its first and second order sensitivities can be evaluated at the j generalised barycenters and at the $M + 1 + j$ vertices. It should be noted that the BDPQA requires only the evaluation of g at the J generalised barycenters after the refinement process of $\hat{\Omega}$. If the refinement strategy requires the evaluation of g during the refinement process, the additional numerical effort may yield $M + 1 + J$ evaluations at the vertices and $J - 1$ evaluations at those generalised barycenters, which are replaced during the refinement process.

- *Geometrical information*
  For a given partition, the coordinates of all supporting points of the inner and outer discretisation are uniquely determined and therefore the distances between the vertices of a subsimplex and their associated general barycenter Λ are easily computed.

The choice of a refinement strategy can be made on the basis of one or several of the three information sets described above. Basic refinement strategies which take only one information set into account will be discussed in the sequel.

- *Strategy based on the probability information*
  A strategy based on the probability information will naturally split the subsimplex with the higher probability weight:

$$\max_j P\left(\hat{\Omega}^j\right).$$ (5.116)

A splitting through the generalised barycenter of the subsimplex ensures an almost equal probability weight in both new subcells of the subsimplex. For a M-dimensional subsimplex there are $M+1$ possibilities of split through the generalised barycenter.

- *Strategy based on the value information*
  Given a partition $\mathcal{P}^j$, g and its first- and second-order sensitivities can be evaluated at the generalised barycenters and vertices. A possible strategy consists in splitting the subsimplex in which g displays the greater variation:

$$\max_j \sum_{m=1}^{M+1} \mid g[v_m^j] - g[\Lambda^j] \mid .$$ (5.117)

The choice of the edge to split can be made in similar way by computing the variation of the value function on the vertices of each edge with respect to the corresponding generalised barycenter. However, high variability of the value function in a subspace of $\hat{\Omega}$ does not necessarily imply low accuracy of the quadratic approximation. An alternative consists in analysing the variability of the second order information, the Hessian matrix, at the vertices with respect to the corresponding generalised barycenters. This is a rather non-trivial task for a large M.

- *Strategy based on the geometrical information*
  Since the quadratic approximation tends to perform worse when the distance between a realisation of $w(T)$ and the supporting point of the approximation increases, a possible choice is to split the subsimplex with the highest total distance between the vertices and their corresponding generalised barycenter:

$$\max_j \sum_{m=1}^{M+1} \parallel v_m^j - \Lambda^j \parallel .$$ (5.118)

The edge to be split is the one with the more distant vertices with respect to their generalised barycenter and the split is made in the middle of the edge.

5.4.6 Numerical Example

The barycentric discretisation of the risk factor space as well as the approximation of the risk profile is illustrated in the sequel with a simple two-dimensional example. The refinement strategy is based on the probability information, the subsimplex with the highest probability weight is split through the generalised barycenter. As this strategy offers $M+1$ possibilities, the longest edge of the subsimplex will be split.

Barycentric Discretisation Let the support of P be following sample of size 10 from a two-dimensional normal distribution $\mathcal{N}(0, I)$, where I is the identity matrix:

$$w^1 = \begin{pmatrix} +1.2800 \\ +1.1299 \end{pmatrix}, w^2 = \begin{pmatrix} +1.6245 \\ +2.5217 \end{pmatrix}, w^3 = \begin{pmatrix} +0.9146 \\ +0.8654 \end{pmatrix}, w^4 = \begin{pmatrix} -1.0699 \\ +0.2029 \end{pmatrix},$$

$$w^5 = \begin{pmatrix} +0.0831 \\ -0.1829 \end{pmatrix}, w^6 = \begin{pmatrix} +0.5779 \\ -0.3888 \end{pmatrix}, w^7 = \begin{pmatrix} +0.7423 \\ -0.7045 \end{pmatrix}, w^8 = \begin{pmatrix} -2.0491 \\ -0.0352 \end{pmatrix},$$

$$w^9 = \begin{pmatrix} +0.4333 \\ -1.2734 \end{pmatrix}, w^{10} = \begin{pmatrix} -0.8202 \\ -1.1998. \end{pmatrix}$$

The vertices of the simplex covering the sample have following coordinates (see Sect. 5.4.2 and Fig. 5.16):

$$A = \begin{pmatrix} 0.000 \\ 8.000 \end{pmatrix}, \qquad B = \begin{pmatrix} -6.9282 \\ -4.000 \end{pmatrix}, \qquad C = \begin{pmatrix} 6.9282 \\ -4.000 \end{pmatrix},$$

the coordinates of the generalised barycenter are computed as follows:

$$\Lambda^{ABC} = \int_{\hat{\Omega}} w d\hat{P} = \frac{1}{10} \sum_{n=1}^{10} w^n = \begin{pmatrix} 0.1716 \\ 0.0935 \end{pmatrix},$$

and the probabilities associated with each vertices are given with:

$$\lambda(\Lambda) = V^{-1} \begin{pmatrix} 1 \\ \Lambda \end{pmatrix} = \begin{pmatrix} 1 & 1 & 1 \\ -6.9282 & 6.9282 & 0.000 \\ -4.000 & -4.000 & 8.000 \end{pmatrix}^{-1} \begin{pmatrix} 1 \\ 0.1716 \\ 0.0935 \end{pmatrix} = \begin{pmatrix} 0.317 \\ 0.342 \\ 0.341 \end{pmatrix}.$$

First Refinement The refinement strategy consists in splitting the longest edge. Since both $[AB]$ and $[AC]$ have the same length, $[AB]$ will be chosen. The new vertex, D, is clearly the intersect of the lines passing through $[AB]$

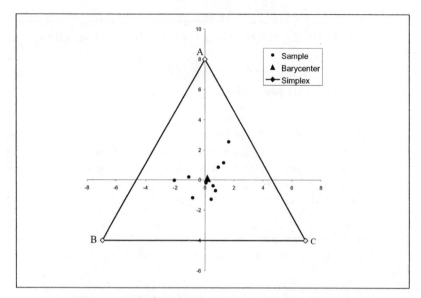

Figure 5.16. Simplicial coverage without refinement

and $[C\Lambda^{ABC}]$. It is easily shown that the equation of the line passing through $[AB]$, respectively $[C\Lambda^{ABC}]$, is:

$$y = 8 + \left(\frac{12}{6.9282}\right) x,$$

respectively:

$$y = -4 - 6.9282 \left(\frac{-4 - 0.0935}{6.9282 - 0.1716}\right) + \left(\frac{-4 - 0.0935}{6.9282 - 0.1716}\right) x,$$

and the coordinates of the new vertex are therefore:

$$D = \begin{pmatrix} -3.3374 \\ 2.2195 \end{pmatrix}.$$

As Fig. 5.17 shows, the partition of $\hat{\Omega}$ is obviously the set $\{\hat{\Omega}^{DAC}, \hat{\Omega}^{BDC}\}$. The probability weights of $\hat{\Omega}^{DAC}$, respectively of $\hat{\Omega}^{BDC}$, are clearly equal to 3/10, respectively to 7/10, which corresponds to the number of realisations of $w(T)$ located in each of the corresponding subsimplex divided by the size of the sample since each realisation has equal probability weight. Each point in $\hat{\Omega}$ is located in a subsimplex if, and only if, all its barycentric weights with respect to this subsimplex are positive.

The generalised barycenters have following coordinates:

$$\Lambda^{DAC} = \int_{\hat{\Omega}^{DAC}} w(T)d\hat{P} = \frac{1}{3}\sum_{n=1}^{3} w^n = \begin{pmatrix} 1.2730 \\ 1.5056 \end{pmatrix},$$

$$\Lambda^{BDC} = \int_{\hat{\Omega}^{BDC}} w(T)d\hat{P} = \frac{1}{7}\sum_{n=4}^{10} w^n = \begin{pmatrix} -0.3003 \\ -0.5116 \end{pmatrix},$$

and the probability weights of the vertices are given with:

$$\lambda(\Lambda^{DAC}) = \hat{\Omega}^{DAC} \left(V^{DAC}\right)^{-1} \begin{pmatrix} 1 \\ \Lambda^{DAC} \end{pmatrix}$$

$$= 0.3 \begin{pmatrix} 1 & 1 & 1 \\ -3.3374 & 0.000 & 6.9282 \\ 2.2195 & 8.000 & -4.000 \end{pmatrix}^{-1} \begin{pmatrix} 1 \\ 1.2730 \\ 1.5056 \end{pmatrix} = \begin{pmatrix} 0.111 \\ 0.080 \\ 0.109 \end{pmatrix},$$

$$\lambda(\Lambda^{BDC}) = \hat{\Omega}^{BDC} \left(V^{BDC}\right)^{-1} \begin{pmatrix} 1 \\ \Lambda^{BDC} \end{pmatrix}$$

$$= 0.7 \begin{pmatrix} 1 & 1 & 1 \\ -6.9282 & -3.3374 & 6.9282 \\ -4.000 & 2.2195 & -4.000 \end{pmatrix}^{-1} \begin{pmatrix} 1 \\ -0.3003 \\ -0.5116 \end{pmatrix} = \begin{pmatrix} 0.074 \\ 0.393 \\ 0.233 \end{pmatrix}.$$

As the vertices C and D are shared by both subsimplices, their associated probability weights are $0.109+0.233 = 0.34$, respectively $0.11+0.393 = 0.504$.

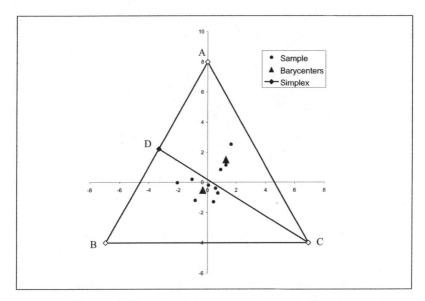

Figure 5.17. Simplicial coverage after one refinement

Second Refinement The above statements are repeated for the second refinement. Additionally, the choice of the subsimplex to be split has to be made. The subsimplex with the highest probability weight, $\hat{\Omega}^{BDC}$, is split on its longest edge $[BC]$ at the point E with (see Fig. 5.18):

$$E = \begin{pmatrix} -3.3374 \\ 2.2195 \end{pmatrix}.$$

The generalised barycenter Λ^{BDC} is replaced by Λ^{DEC} and Λ^{BDE} with following coordinates:

$$\Lambda^{DEC} = \int_{\hat{\Omega}^{DEC}} w(T)d\hat{P} = \frac{1}{4}\sum_{n=4}^{7} w^n = \begin{pmatrix} 0.0833 \\ -0.2683 \end{pmatrix},$$

$$\Lambda^{BDE} = \int_{\hat{\Omega}^{BDE}} w(T)d\hat{P} = \frac{1}{3}\sum_{n=8}^{10} w^n = \begin{pmatrix} -0.8120 \\ -0.8362 \end{pmatrix}.$$

For both new subcells the probability weights of the vertices are calculated as above for the first refinement.

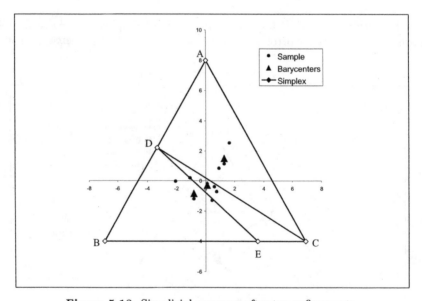

Figure 5.18. Simplicial coverage after two refinements

Approximation of the Portfolio Distribution Let's examine a portfolio, consisting of one call option on each risk factor with same strike price 60 and maturity three months. The risk-free interest rate is 5% and a realisation of the risk factors is given with:

$$50e^{w_m 0.25\sqrt{\frac{1}{4}}},$$

where $m \in \{1, 2\}$. Hence $w(0) = (50, 50)$, the volatility yields 25% in each dimension and the horizon is three months. The profit and loss distribution, $F_{\bar{g}}$, where g is evaluated with the Black-Scholes formula is given with:

$$F_{\bar{g}} \begin{cases} \begin{array}{cc} \text{Percentile} & \text{Value} \\ 10\% & -0.5015 \\ 20\% & -0.2930 \\ 30\% & -0.1072 \\ 40\% & -0.0390 \\ 50\% & 0.1321 \\ 60\% & 0.4388 \\ 70\% & 0.6703 \\ 80\% & 2.4407 \\ 90\% & 4.2645 \\ 100\% & 13.3182 \end{array} \end{cases}$$

For illustrative purposes, the results of the approximation of $F_{\bar{g}}^{\nabla}$ without $(F_{\bar{g}}^{\nabla,0})$ and after the first $(F_{\bar{g}}^{\nabla,1})$, respectively the second refinement $(F_{\bar{g}}^{\nabla,2})$ are summarised in table 5.6. $F_{\bar{g}}^{\nabla,0}$ yields results that are very similar to the Delta-Gamma approximation (or Delta-Gamma Monte Carlo) as the generalised barycenter Λ^{ABC} is located at the center of the distribution. Without sampling error the coordinates of Λ^{ABC} would be $(0,0)^{\top}$. The range of the percentage error diminishes rapidly with the number of refinements from $[-1239.54\%, 286.52\%]$ to $[-11.65\%, 33.32\%]$. For 8 out of 10 observations the percentage error is below 10%. The mean of the absolute values of the percentage error is 218% for $(F_{\bar{g}}^{\nabla,0})$, 52.8% for $(F_{\bar{g}}^{\nabla,1})$ and 7.12% for $(F_{\bar{g}}^{\nabla,2})$. Figure 5.19 illustrates the decrease in error in absolute terms. It is worth noting, that the accuracy of the approximation has been increased for all percentiles of $F_{\bar{g}}$. It can easily be verified, that the approximation performs at best for those realisations of $w(T)$ which are in the neighbourhood of their associated generalised barycenters.

5.5 Benchmarking the BDPQA

In this section the BDPQA, with $J = 25$, is compared with the global quadratic approximation for the estimation of risk measures of non-linear option portfolios. The characteristics of the portfolios are described in Sect. 1.6. Since the Delta-Gamma approximation is only applicable for normal distributed risk factors, the global quadratic approximation is obtained with

$F_{\bar{g}}^{\nabla,0}$

Percentile	Approximation	Absolute error	Percentage error
10%	-0.0345	-0.4670	-93.12%
20%	0.0051	-0.2982	-101.76%
30%	0.2100	-0.3172	-295.94%
40%	0.4450	-0.4840	-1239.54%
50%	0.5107	-0.3786	286.52%
60%	0.7174	-0.2786	63.51%
70%	1.0332	-0.3628	54.13%
80%	2.2329	0.2079	-8.52%
90%	3.6828	0.5816	-13.64%
100%	10.2120	3.1062	-23.32%

$F_{\bar{g}}^{\nabla,1}$

Percentile	Approximation	Absolute error	Percentage error
10%	-0.4518	-0.0497	-9.91%
20%	-0.1130	-0.1800	-61.42%
30%	-0.0548	-0.0524	-48.88%
40%	0.0913	-0.1303	-333.76%
50%	0.1309	0.0013	-0.95%
60%	0.2947	0.1441	-32.83%
70%	0.4229	0.2474	-36.91%
80%	2.4194	0.0213	-0.87%
90%	4.2584	0.0060	-0.14%
100%	13.6237	-0.3055	2.29%

$F_{\bar{g}}^{\nabla,2}$

Percentile	Approximation	Absolute error	Percentage error
10%	-0.4985	-0.0031	-0.61%
20%	-0.2719	-0.0211	-7.21%
30%	-0.1429	0.0357	33.32%
40%	-0.0392	0.0001	0.28%
50%	0.1167	0.0154	-11.65%
60%	0.4105	0.0282	-6.44%
70%	0.6141	0.0562	-8.39%
80%	2.4194	0.0213	-0.87%
90%	4.2584	0.0060	-0.14%
100%	13.6237	-0.3055	2.29%

Table 5.6. Approximations of $F_{\bar{g}}$

the Delta-Gamma Monte Carlo method. As a result, the discrepancies in the estimations do not stem from differences in the risk factor modelling but exclusively from differences in the risk profile approximation.

For each portfolio, the support of P is a sample of 10'000 realisations of $w(T)$ drawn at pseudo-random. The α percentiles, with $\alpha = 1\%$, 3%, and 5%, of the profit and loss distributions, i.e. VaR, denoted by $v_{-,\alpha}$, as well as the corresponding TailVaR numbers, denoted by $tv_{-,\alpha}$ for the short holdings

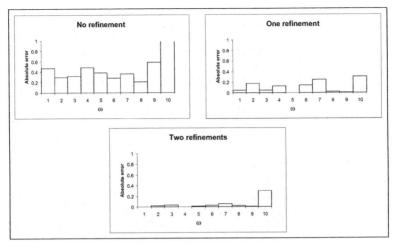

Figure 5.19. Absolute errors of the approximation

are listed in the tables below. The benchmark is obtained by full valuation. $v_{-,\alpha}^{\nabla}$, respectively $tv_{-,\alpha}^{\nabla}$, denotes the VaR, respectively TailVaR, estimates of the BDPQA. $v_{-,\alpha}^{\Delta-\Gamma}$ and $tv_{-,\alpha}^{\Delta-\Gamma}$ stand for the corresponding statistics of the Delta-Gamma Monte Carlo method. Both VaR and TailVaR are measures focussing on tail events and are therefore difficult to estimate.

Tables 5.7 to 5.9 summarise the results for normally distributed risk factors and a holding period of 10 days. The next three tables (5.10 to 5.12) display the same statistics for the portfolios of environment I under the alternative assumption of Student-t distributed log-returns and a holding period of two weeks. Finally, the effect of the holding period on the goodness of the approximations are illustrated in tables 5.13 to 5.15 for the FX-portfolio of environment II.

Normally Distributed Log-Returns The BDPQA yields significantly better results than the quadratic approximation at all α levels and for all test portfolios. The accuracy of both approximations is relatively insensitive with respect to the level α. The evaluation of portfolio PF_3^I, an inverse butterfly spread whose risk profile exhibits sudden changes in slope, is still challenging for both approximations.

Student-t Distributed Log-Returns As for normally distributed log-returns, both methods produce conservative risk estimates for all test portfolios. The BDPQA remains significantly superior to the global quadratic approximation for all portfolios and at all α levels. Moreover, the global quadratic approximation is more sensitive with respect to the level of α than the BDPQA. Note the important impact of an alternative risk factor modelling on the VaR and TailVaR numbers.

	TailVaR			VaR		
$\alpha = 1\%$	$tv_{-,\alpha}^{\nabla}$	$tv_{-,\alpha}$	Error	$v_{-,\alpha}^{\nabla}$	$v_{-,\alpha}$	Error
PF_1^I	4025.26	3866.07	4.12 %	3'418.88	3'278.50	4.28%
PF_2^I	3741.01	3513.02	6.49 %	3'191.52	3'037.48	5.07%
PF_3^I	3745.93	3215.18	16.51 %	3'053.18	2'690.03	13.50%
PF_4^I	1660.59	1064.84	55.95 %	1'405.06	961.35	46.15%
PF^{II}	2734.35	2591.32	5.52 %	2'279.31	2'172.83	4.90%
$\alpha = 3\%$						
PF_1^I	3348.38	3216.33	4.11 %	2'687.45	2'580.19	4.16%
PF_2^I	3092.73	2944.51	5.03 %	2'471.65	2'397.49	3.09%
PF_3^I	3042.68	2664.22	14.21 %	2'429.67	2'166.96	12.12%
PF_4^I	1377.83	940.70	46.47 %	1'124.19	823.20	36.56%
PF^{II}	2216.58	2117.81	4.66 %	1'734.69	1'658.86	4.57%
$\alpha = 5\%$						
PF_1^I	3006.36	2892.30	3.94 %	2'322.62	2'249.73	3.24%
PF_2^I	2772.16	2653.93	4.45 %	2'125.68	2'060.27	3.17%
PF_3^I	2723.96	2397.80	13.60 %	2'081.67	1'870.90	11.27%
PF_4^I	1248.57	876.91	42.38 %	995.64	744.91	33.66%
PF^{II}	1967.60	1880.88	4.61 %	1'484.19	1'413.90	4.97%
Horizon: 10 days						

Table 5.7. BDPQA for normally distributed log-returns

Different Holding Periods Tables 5.12 to 5.15 illustrate the impact of the holding period on the accuracy of the approximation for portfolio PF^{II}. The dispersion of the risk factors increases with the passage of time. This is reflected by the increasing Value-at-Risk and TailVaR numbers, which are particularly consequent at the 1% level. The quadratic approximation worsens dramatically as the holding period widens. On the other hand, the BDPQA shows a great stability, yielding even better results for longer horizons.

Figure 5.20 is the graphical depiction of the density function of $\bar{g}[w(T)]$ and its approximations for a horizon of 90 days. Clearly, the distribution based on the quadratic approximation overestimates the probability of high loss. The distribution obtained with the BDPQA fits the real distribution much better over the whole domain of $\bar{g}[w(T)]$. Figure 5.21 focusses on the goodness of both approximations in the loss part of $\bar{g}[w(T)]$ and shows the superiority of the BDPQA over the global approximation for the estimation of tails events.

Behaviour of the BDPQA with Increasing J As stated above, for the BDPQA the number of full evaluations of $\bar{g}[w(T)]$, as well as the number of computations of their associated Gradient vectors and Hessian matrixes, increases linearly with the number J of refinements. A small J ensures therefore the numerical tractability of the approximation. Figure 5.22 illustrates the

	TailVaR			VaR		
$\alpha = 1\%$	$tv^{\Delta-\Gamma}_{-,\alpha}$	$tv_{-,\alpha}$	Error	$v^{\Delta-\Gamma}_{-,\alpha}$	$v_{-,\alpha}$	Error
PF_1^I	4639.87	3866.07	20.02 %	3'808.95	3'278.50	16.18%
PF_2^I	3990.07	3513.02	13.58 %	3'372.21	3'037.48	11.02%
PF_3^I	4867.18	3215.18	51.38 %	3'798.43	2'690.03	41.20%
PF_4^I	2702.66	1064.84	153.81 %	2'129.91	961.35	121.55%
PF^{II}	4480.30	2591.32	72.90 %	3'475.08	2'172.83	59.93%
$\alpha = 3\%$						
PF_1^I	3743.33	3216.33	16.39 %	2'892.20	2'580.19	12.09%
PF_2^I	3270.56	2944.51	11.07 %	2'584.97	2'397.49	7.82%
PF_3^I	3819.55	2664.22	43.36 %	2'904.90	2'166.96	34.05%
PF_4^I	2120.40	940.70	125.41 %	1'613.11	823.20	95.96%
PF^{II}	3401.07	2117.81	60.59 %	2'471.91	1'658.86	49.01%
$\alpha = 5\%$						
PF_1^I	3316.40	2892.30	14.66 %	2'490.63	2'249.73	10.71%
PF_2^I	2916.29	2653.93	9.89 %	2'199.82	2'060.27	6.77%
PF_3^I	3353.06	2397.80	39.84 %	2'473.82	1'870.90	32.23%
PF_4^I	1866.09	876.91	112.80 %	1'381.88	744.91	85.51%
PF^{II}	2911.26	1880.88	54.78 %	1'983.34	1'413.90	40.27%
Horizon: 10 days						

Table 5.8. Quadratic approximation for normally distributed log-returns

Error	$\alpha = 1\%$		$\alpha = 3\%$		$\alpha = 5\%$	
	$tv^{\nabla}_{-,\alpha}$	$tv^{\Delta-\Gamma}_{-,\alpha}$	$tv^{\nabla}_{-,\alpha}$	$tv^{\Delta-\Gamma}_{-,\alpha}$	$tv^{\nabla}_{-,\alpha}$	$tv^{\Delta-\Gamma}_{-,\alpha}$
PF_1^I	4.12 %	20.02 %	4.11 %	16.39 %	3.94 %	14.66 %
PF_2^I	6.49 %	13.58 %	5.03 %	11.07 %	4.45 %	9.89 %
PF_3^I	16.51 %	51.38 %	14.21 %	43.36 %	13.60 %	39.84 %
PF_4^I	55.95 %	153.81 %	46.47 %	125.41 %	42.38 %	112.80 %
PF^{II}	5.52 %	72.90 %	4.66 %	60.59 %	4.61 %	54.78 %
	$v^{\nabla}_{-,\alpha}$	$v^{\Delta-\Gamma}_{-,\alpha}$	$v^{\nabla}_{-,\alpha}$	$v^{\Delta-\Gamma}_{-,\alpha}$	$v^{\nabla}_{-,\alpha}$	$v^{\Delta-\Gamma}_{-,\alpha}$
PF_1^I	4.28%	16.18%	4.16%	12.09%	3.24%	10.71%
PF_2^I	5.07%	11.02%	3.09%	7.82%	3.17%	6.77%
PF_3^I	13.50%	41.20%	12.12%	34.05%	11.27%	32.23%
PF_4^I	46.15%	121.55%	36.56%	95.96%	33.66%	85.51%
PF^{II}	4.90%	59.93%	4.57%	49.01%	4.97%	40.27%
Horizon: 10 days						

Table 5.9. Error for normally distributed log-returns

behaviour of the BDPQA with respect to the number of refinements for the FX-portfolio with an holding period of ten days and is representative for all portfolios that have been tested. Clearly, the first five refinements contribute

$\alpha = 1\%$	TailVaR			VaR		
	$tv^{\nabla}_{-,\alpha}$	$tv_{-,\alpha}$	Error	$v^{\nabla}_{-,\alpha}$	$v_{-,\alpha}$	Error
PF_1^I	5618.87	5264.76	6.73 %	4'149.58	3'914.65	6.00%
PF_2^I	4409.78	4119.24	7.05 %	3'514.55	3'228.55	8.86%
PF_3^I	6011.06	4948.25	21.48 %	4'323.45	3'565.00	21.27%
PF_4^I	2970.59	1355.26	119.19 %	2'168.78	1'231.92	76.05%
$\alpha = 3\%$						
PF_1^I	4213.86	3915.53	7.62 %	3'017.70	2'833.75	6.49%
PF_2^I	3430.67	3230.15	6.21 %	2'579.41	2'440.80	5.68%
PF_3^I	4425.43	3658.66	20.96 %	3'146.63	2'614.51	20.35%
PF_4^I	2196.29	1198.53	83.25 %	1'588.33	1'045.78	51.88%
$\alpha = 5\%$						
PF_1^I	3611.87	3362.51	7.42 %	2'448.87	2'292.72	6.81%
PF_2^I	3002.12	2838.92	5.75 %	2'161.98	2'092.76	3.31%
PF_3^I	3799.77	3159.00	20.28 %	2'623.45	2'244.82	16.87%
PF_4^I	1913.35	1115.45	71.53 %	1'401.75	944.92	48.35%
Horizon: 10 days						

Table 5.10. BDPQA for Student-t distributed log-returns

$\alpha = 1\%$	TailVaR			VaR		
	$tv^{\Delta\Gamma}_{-,\alpha}$	$tv_{-,\alpha}$	Error	$v^{\Delta\Gamma}_{-,\alpha}$	$v_{-,\alpha}$	Error
PF_1^I	7975.41	5264.76	51.49%	4'940.15	3'914.65	26.20%
PF_2^I	5860.60	4119.24	42.27%	4'073.59	3'228.55	26.17%
PF_3^I	10898.54	4948.25	120.25%	6'160.63	3'565.00	72.81%
PF_4^I	7156.91	1355.26	428.08%	3'984.33	1'231.92	223.42%
$\alpha = 3\%$						
PF_1^I	5319.06	3915.53	35.84%	3'403.37	2'833.75	20.10%
PF_2^I	4173.63	3230.15	29.21%	2'839.57	2'440.80	16.34%
PF_3^I	6876.22	3658.66	87.94%	3'976.76	2'614.51	52.10%
PF_4^I	4516.65	1198.53	276.85%	2'620.10	1'045.78	150.54%
$\alpha = 5\%$						
PF_1^I	4384.98	3362.51	30.41%	2'657.69	2'292.72	15.92%
PF_2^I	3534.07	2838.92	24.49%	2'346.25	2'092.76	12.11%
PF_3^I	5563.26	3159.00	76.11%	3'248.27	2'244.82	44.70%
PF_4^I	3647.34	1115.45	226.98%	2'143.78	944.92	126.87%
Horizon: 10 days						

Table 5.11. Quadratic approximation for Student-t distributed log-returns

to the greatest part of the increase in accuracy, which ensures the tractability of the method.

Error	$\alpha = 1\%$		$\alpha = 3\%$		$\alpha = 5\%$	
	$tv_{-,\alpha}^{\nabla}$	$tv_{-,\alpha}^{\Delta-\Gamma}$	$tv_{-,\alpha}^{\nabla}$	$tv_{-,\alpha}^{\Delta-\Gamma}$	$tv_{-,\alpha}^{\nabla}$	$tv_{-,\alpha}^{\Delta-\Gamma}$
PF_1^I	6.73 %	51.49 %	7.62 %	35.84 %	7.42 %	30.41 %
PF_2^I	7.05 %	42.27 %	6.21 %	29.21 %	5.75 %	24.49 %
PF_3^I	21.48 %	120.25 %	20.96 %	87.94 %	20.28 %	76.11 %
PF_4^I	119.19 %	428.08 %	83.25 %	276.85 %	71.53 %	226.98 %
	$v_{-,\alpha}^{\nabla}$	$v_{-,\alpha}^{\Delta-\Gamma}$	$v_{-,\alpha}^{\nabla}$	$v_{-,\alpha}^{\Delta-\Gamma}$	$v_{-,\alpha}^{\nabla}$	$v_{-,\alpha}^{\Delta-\Gamma}$
PF_1^I	6.00%	26.20%	6.49%	20.10%	6.81%	15.92%
PF_2^I	8.86%	26.17%	5.68%	16.34%	3.31%	12.11%
PF_3^I	21.27%	72.81%	20.35%	52.10%	16.87%	44.70%
PF_4^I	76.05%	223.42%	51.88%	150.54%	48.35%	126.87%
Horizon: 10 days						

Table 5.12. Error for Student-t distributed log-returns

Horizon	TailVaR			VaR		
$\alpha = 1\%$	$tv_{-,\alpha}^{\nabla}$	$tv_{-,\alpha}$	Error	$v_{-,\alpha}^{\nabla}$	$v_{-,\alpha}$	Error
2 weeks	2'734	2'591	5.52 %	2'279	2'172	4.90%
3 months	13'048	12'839	1.63 %	10'612	10'391	2.13%
1 year	62'276	63'644	-2.15 %	44'383	45'099	-1.59%
$\alpha = 3\%$						
2 weeks	2'216	2'117	4.66%	1'734	1'658	4.57%
3 months	10'256	10'089	1.65%	7'610	7'537	0.97%
1 year	44'063	44'949	-1.97%	28'062	28'018	0.16%
$\alpha = 5\%$						
2 weeks	1'967	1'880	4.61%	1'484	1'413	4.97%
3 months	8'926	8'759	1.91%	6'364	6'210	2.49%
1 year	36'339	37'018	-1.83%	22'082	21'962	0.55%
Distribution of log-returns: normal						

Table 5.13. BDPQA for profile $\bar{g}^{II}$ with various horizons

5.5.1 The Choice of the Holding Period

As shown above, the accuracy of the analytical approximations and therefore their appropriateness depends on the considered holding period. The choice of an appropriate holding period is influenced by two key factors: the interval of time between asset allocation and/or hedging decisions and the period required to liquidate a position. For trading books, which exhibit a very high turnover and generally contain instruments traded on very liquid markets, a holding period of one to ten days appears to be appropriate. Note, however, that even the more active markets can become fairly illiquid in the event of a major crisis. For banking books, positions of non-financial institutions,

Horizon	TailVaR			VaR		
$\alpha = 1\%$	$tv_{-,\alpha}^{\Delta-\Gamma}$	$tv_{-,\alpha}$	Error	$v_{-,\alpha}^{\Delta-\Gamma}$	$v_{-,\alpha}$	Error
2 weeks	4'480	2'591	72.90%	3'475	2'172	59.93%
3 months	59'249	12'839	361.46%	38'575	10'391	271.24%
1 year	1'381'319	63'644	2'070.37 %	569'322	45'099	1162.38%
$\alpha = 3\%$						
2 weeks	3'401	2'117	60.59%	2'471	1'658	49.01%
3 months	38'869	10'089	285.25 %	21'351	7'537	183.27%
1 year	713'578	44'949	1'487.51 %	242'831	28'018	766.68%
$\alpha = 5\%$						
2 weeks	2'911	1'880	54.78 %	1'983	1'413	40.27%
3 months	30'727	8'759	250.80 %	16'107	6'210	159.38%
1 year	502'924	37'018	1'258.56%	144'160	21'962	556.39%
Distribution of log-returns: normal						

Table 5.14. Quadratic approximation for profile $\bar{g}^{II}$ with various horizons

Error	$\alpha = 1\%$		$\alpha = 3\%$		$\alpha = 5\%$	
Horizon	$tv_{-,\alpha}^{\nabla}$	$tv_{-,\alpha}^{\Delta-\Gamma}$	$tv_{-,\alpha}^{\nabla}$	$tv_{-,\alpha}^{\Delta-\Gamma}$	$tv_{-,\alpha}^{\nabla}$	$tv_{-,\alpha}^{\Delta-\Gamma}$
2 weeks	5.52 %	72.90 %	4.66 %	60.59 %	4.61 %	54.78 %
3 months	1.63 %	361.46 %	1.65 %	285.25 %	1.91 %	250.80 %
1 year	-2.15 %	2'070 %	-1.97 %	1'488 %	-1.83 %	1'259 %
Horizon	$v_{-,\alpha}^{\nabla}$	$v_{-,\alpha}^{\Delta-\Gamma}$	$v_{-,\alpha}^{\nabla}$	$v_{-,\alpha}^{\Delta-\Gamma}$	$v_{-,\alpha}^{\nabla}$	$v_{-,\alpha}^{\Delta-\Gamma}$
2 weeks	4.90%	59.93%	4.57%	49.01%	4.97%	40.27%
3 months	2.13%	271.24%	0.97%	183.27%	2.49%	159.38%
1 year	-1.59%	1'162%	0.16%	766.68%	0.55%	556.39%
Distribution of log-returns: normal						

Table 5.15. Error of for $\bar{g}^{II}$ with various horizons

or institutional investors the risk management horizon varies typically from months to years. Furthermore, tailor-made over-the-counter instruments exhibit no liquidity at all and their risk exposure is often difficult to hedge. This suggests a holding period equivalent to the remaining time to maturity of the considered instruments. Finally, the assessment of credit risk requires the measurement of the market exposure over the whole life of the financial instruments of interest (see Sect. 7.2). The corresponding horizon may yield several years.

5.6 Summary

Considering high-dimensional, non-linear portfolios, the estimation of the portfolio distribution F_g typically involves a trade-off between accuracy and

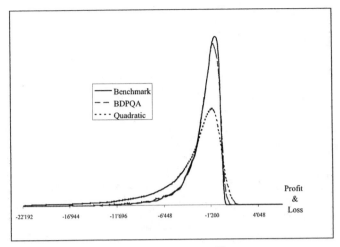

Figure 5.20. Density function of $\bar{g}^{II}$, horizon 90 days

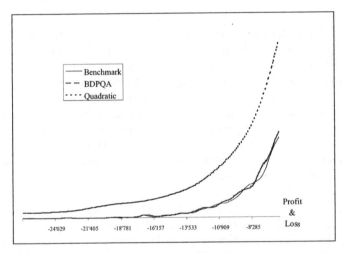

Figure 5.21. Density function of $\bar{g}^{II}$, left tail, horizon 90 days

numerical tractability. The analytical approximations necessitate a single full valuation of the portfolio profile as well as the computation of M partial derivatives of first order and, for the Delta-Gamma approximation, $\frac{M(M_2+1)}{2}$ derivatives of second order. The analytical approximations do not allow for an adequate modelling of the risk factor dynamics and fail to capture the whole non-linearity of optioned portfolios. The accuracy of their estimates decreases very fast with an increasing holding period.

On the other hand, the Monte Carlo methods allow a realistic modelling of the risk factor distributions and generate estimates, at least for the (ide-

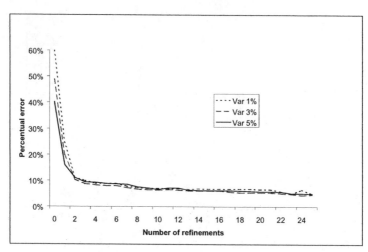

Figure 5.22. Behaviour of the BDPQA

alised) pseudo-random method, which converge to F_g or to the true values of risk measures derived from F_g and which are independent of the dimensionality M of the risk factor space. The empirical results suggest that the quasi-random scheme is superior to its pseudo-random equivalent especially for a moderate M. However, the theoretical assessment of the quasi-random method is not that easy since scenarios generated by low-discrepancy sequences are no longer independent. For both the pseudo- and quasi-random Monte Carlo methods accurate estimates require thousand calls of the valuation function and may become impractical for large portfolios containing instruments with complex pay-offs.

The BDPQA, like the Monte Carlo methods, poses no assumptions on the dynamics of the risk factors. While it allows large samples of size N^1, the number of full valuations is limited to J. The numerical results show that the BDPQA outperforms the Delta-Gamma approximation consistently and that it provides accurate estimates for most of the tested portfolios. Moreover, the estimates are robust with respect to the holding period.

6. Sample Estimation of Risk Measures

6.1 Introduction

As shown in the last chapter, the analytical methods used for the approximation of the portfolio distribution make strong assumptions on the dynamics of the risk factors and use global Taylor expansion-based approximations of the pricing functions. More versatile methods like the BDPQA and the Monte Carlo simulation which allow for a more realistic modelling of risk factors and for the use of exact pricing formulas are based on a discretisation of the risk factor space. As a result, the approximation of the portfolio distribution F_g is not given analytically or numerically but in form of a sample of size N from F_g. The evaluation of a risk measure derived from the portfolio distribution bases therefore on the generated sample. Since statistics located in the tails of the distribution are notoriously hard to estimate, the risk measure considered in this chapter is Value-at-Risk at the levels 1%, 3%, and 5%. Portfolio PF^{II} has been chosen for the analysis with a holding period of 90 days.

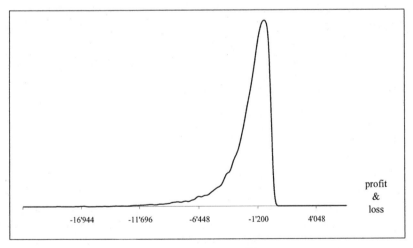

Figure 6.1. Density function of $\bar{g}^{II}$, horizon 90 days

Figure 6.1 is the graphical depiction of the density function of its profit and loss and shows that the distribution is heavily skewed and long tailed with the consequence that estimators of low quantiles become very sensitive to the variability of the sample. The first part of the chapter deals with the relationship between *order statistics* and sample quantiles and analyse different quantile estimators based on single order statistics or simple combinations of adjacent order statistics. The second part considers weighted averages of various order statistics as a possible enhancement for the quantile estimation. Due to the averaging, the estimators exhibit less variability and hence produce more reliable estimates of Value-at-Risk particularly when the size of the sample is limited. The relative performance of several estimators for different small samples is analysed at the end of the chapter.

6.2 Order Statistics

The treatment of this section follows in part the conceptual layouts presented in Gibbons and Chakraborti (1992) and Reiss (1989).

Let $X_1, X_2, \cdots, X_N$ be a random sample from a population with common distribution function F_X. It will be further assumed that F_X is continuous and admits a density f_X. Consider the ordered values:

$$X_{1:N} \leq X_{2:N} \leq \cdots \leq X_{N:N}. \tag{6.1}$$

$X_{r:N}$ for $1 \leq r \leq N$ is called the rth *order statistic* and the random vector $(X_{1:N}, X_{2:N}, \cdots, X_{N:N})$ is the order statistic. If F_X is absolutely continuous, ties in the sample occur with probability 0 and there exists a unique ordered arrangement:

$$X_{1:N} < X_{2:N} < \cdots < X_{N:N}. \tag{6.2}$$

6.2.1 Distribution of Order Statistics

As the X_i are identically distributed variables the joint probability density function is:

$$f_{X_1,\cdots,X_N}(x_1,\cdots,x_N) = \prod_{n=1}^{N} f_X(x_n). \tag{6.3}$$

The joint density of the ordered sample can be easily derived from the joint probability density function f_X by using the method of Jacobians of transformations. The joint distribution of N functions Y_n of $X_1, \cdots, X_N$ when $Y_1 = Y_1(X_1, \cdots, X_N) \cdots Y_N = Y_1(X_1, \cdots, X_N)$ and the transformation from $(X_1, \cdots, X_N)$ to $(Y_1, \cdots, Y_N)$ is one to one and given by (see Johnson et al. (1994)):

$$f_{Y_1,\cdots,Y_N}(y_1,\cdots,y_N) = f_{X_1,\cdots,X_N}(x_{1(y)},\cdots,x_{N(y)}) \left| \frac{\partial\left(x_{1(y)},\cdots,x_{N(y)}\right)}{\partial\left(y_1,\cdots,y_N\right)} \right|,$$

$$(6.4)$$

where $x_n(y)$ means that x_n is expressed in terms of $y_1,\cdots,y_N$, i.e. the unique inverse transformation and $\frac{\partial(x_1(y),\cdots,x_N(y))}{(y_1,\cdots,y_N)} = J$ is the Jacobian of $(x_1,..,x_N)$ with respect to $(y_1,..,y_N)$, i.e. the determinant of the $N \times N$ matrix of the partial derivatives corresponding to the inverse transformation.

Consider following transformation:

$$Y_n = X_{n:N} \qquad n = 1,\cdots,N. \tag{6.5}$$

The transformation is not one to one. Every ordered collection could have come from $N!$ different samples since the original variables can be arranged in $N!$ ways (in case F_X is absolutely continuous, i.e. there is no tie). For each of these $N!$ permutations Y_n equals exactly one of the X_n and reciprocally. Each row and column of the matrix of the partial derivatives contains therefore $(N-1)$ entries with value zero and one entry with value one. Thus the matrix is orthogonal and its determinant is ± 1. The absolute value of the Jacobian is therefore one and:

$$f_{X_{1:N},\cdots,X_{N:N}}(y_1,\cdots,y_n) = \sum_{\substack{number \\ of \\ permutations}} \prod_{n=1}^{N} f_{X_n,\cdots,X_N}(x_n) \mid J \mid,$$

$$(6.6)$$

$$= N! \prod_{n=1}^{N} f_X(y_n).$$

with $y_1 < \cdots < y_N$ and $X = (X_n,\cdots,X_N)$.

The marginal distribution of the rth order statistic is found by successive integration of the joint frequency function over the other order statistics:

$$f_{X_{r:N}} = N! f_X(y_r) \int\limits_{-\infty}^{y_r} \cdots \int\limits_{-\infty}^{y_2} \int\limits_{y_r}^{\infty} \cdots \int\limits_{y_{N-1}}^{\infty} \prod_{\substack{n=1 \\ n \neq r}}^{N} f_X(y_n) dy_N \cdots dy_{r+1} dy_1 \cdots dy_{r-1}$$

$$= N! f_X(y_r) \int\limits_{-\infty}^{y_r} \cdots \int\limits_{-\infty}^{y_2} \int\limits_{y_r}^{\infty} \cdots \int\limits_{y_{N-2}}^{\infty} [1 - F_X(y_{N-1})] f_X(y_{N-1})$$

$$\times \prod_{\substack{n=1 \\ n \neq r}}^{(N-2)} f_X(y_n) dy_{N-1} \cdots dy_{r+1} dy_1 \cdots dy_{r-1}$$

$$= N! f_X(y_r) \int\limits_{-\infty}^{y_r} \cdots \int\limits_{-\infty}^{y_2} \int\limits_{y_r}^{\infty} \cdots \int\limits_{y_{N-3}}^{\infty} \frac{[1 - F_X(y_{N-2})]^2}{2 \cdots 1} f_X(y_{N-2})$$

$$\times \prod_{\substack{n=1 \\ n \neq r}}^{(N-3)} f_X(y_n) dy_{N-2} \cdots dy_{r+1} dy_1 \cdots dy_{r-1}$$

. . . .

$$= \frac{N!}{(r-1)!(N-r)!} [F_X(y_r)]^{r-1} [1 - F_X(y_r)]^{N-r} f_X(y_r).$$

$$(6.7)$$

In the case of the uniform distribution the marginal distribution of the rth order statistic is given with:

$$f_{X_{r:N}} = \frac{N!}{(r-1)!(N-r)!} x^{r-1} (1-x)^{N-r}. \tag{6.8}$$

Let $I_{x \in (-\infty, s]}$ denote the indicator function of the interval $(-\infty, s]$, i.e. $I_B = 1$ if $x \in B$ and $= 0$ if $x \notin B$. The frequency of the data may be written $\sum_{n=1}^{N} I_{x \in (-\infty, s]}$. Following holds for the rth order statistic:

$$\{X_{r:N} \leq s\} \text{ iff } \left\{ \sum_{n=1}^{N} I_{X_n \in (-\infty, s]} \geq r \right\}, \tag{6.9}$$

and thus:

$$P\{X_{r:N} \leq s\} = P\left\{ \sum_{n=1}^{N} I_{X_n \in (-\infty, s]} \geq r \right\}. \tag{6.10}$$

As $I_{X_n \in (-\infty, s]}$ is a binomial random variable:

$$P\{X_{r:N} \leq s\} = \sum_{n=r}^{N} \frac{N!}{n!(N-n)!} [F_X(s)]^n [1 - F_X(s)]^{N-n}. \tag{6.11}$$

The relation between the order statistics and the sample (or empirical) distribution function $F_N(s)$, which is defined as follows, is immediate:

$$F_N(s) = \begin{cases} 0 & \text{if } s < X_{1:N}, \\ \frac{r}{N} & \text{if } X_{r:N} \le s < X_{r+1:N} \\ 1 & \text{if } s \ge X_{N:N}. \end{cases} \qquad \text{for } r = 1, \cdots, N-1, \qquad (6.12)$$

From the definition:

$$P\left\{ F_N(s) \ge \frac{r}{N} \right\} = P\left\{ \sum_{n=1}^{N} I_{X_n \in (-\infty, s]} \ge r \right\}, \qquad (6.13)$$

and thus:

$$P\left\{ F_N(s) \ge \frac{r}{N} \right\} = P\{X_{r:N} \le s\}. \qquad (6.14)$$

6.2.2 Moments of Order Statistics

The *quantile function* $Q_X(\alpha)$ is characterised by:

$$Q_X(\alpha) = q_\alpha = F_X^{-1}(\alpha) = \inf\{s : F_X(s) \ge \alpha\}, \qquad (6.15)$$

where $0 < \alpha < 1$.

If F_X is continuous and strictly increasing, F_X^{-1} corresponds to the mathematical definition of the inverse of F_X and $Q_X(\alpha)$ is equal to the unique solution t of the equation $F_X(t) = \alpha$.

The kth moment about the origin of the rth order statistic from F_X is:

$$E\left[X_{r:N}^k \right] = \int_{-\infty}^{\infty} y^k f_{X_{r:N}}(y) dy$$

$$= \frac{N!}{(r-1)!(N-r)!} \int_{-\infty}^{\infty} y^k \left[F_X(y_r) \right]^{r-1} \left[1 - F_X(y_r) \right]^{N-r} f_X(y) dy$$

$$= \frac{N!}{(r-1)!(N-r)!} \int_{-\infty}^{\infty} \left[F_X(y_r) \right]^{r-1} \left[1 - F_X(y_r) \right]^{N-r} dF_X(y)$$

$$= \frac{N!}{(r-1)!(N-r)!} \int_0^1 \left[Q_X(u) \right]^k u^{r-1} (1-u)^{N-r} du$$

$$= E\left[Q_X(U) \right]^k , \qquad (6.16)$$

where U has Beta distribution with parameters r and $N - r + 1$.

6.2.3 Confidence Interval for Population Quantiles

When considering a sequence $X_{r:N}$ of order statistics, it should be distinguished between two cases:

- $\lim_{N \to \infty} \frac{r}{N} = \alpha$ with $0 < \alpha < 1$.
- $\lim_{N \to \infty} \frac{r}{N} = \alpha$ with $\alpha = 0$ or $\alpha = 1$, i.e. r, respectively $N - r$, is bounded.

The first case is referred to as a *central sequence of order statistics*, the second one as a sequence of *lower (upper) extremes*. Central and extreme sequences have completely different asymptotical properties. Central order statistics are generally of interest for the inference of quantiles and it can be shown that they are asymptotically normally distributed with (see Manoukian (1986)):

$$\sqrt{N} \frac{X_{r:N} - q_\alpha}{\sqrt{\alpha(1 - \alpha)}} \sim \Phi\left(0, \frac{1}{f_X^2(q_\alpha)}\right) \tag{6.17}$$

for f_X positive and continuous near q_α and $r = [N\alpha] + 1$, with $[N\alpha] = $ the larger integer less or equal $N\alpha$. $\Phi()$ denotes the cumulative distribution function of the standard normal distribution.

In the case of F_X strictly monotone following holds:

$$X_{r:N} \xrightarrow{P} q_\alpha. \tag{6.18}$$

Thus, $X_{r:N}$ is a consistent estimator of the αth-quantile of F_X and therefore the sample or empirical quantile function is defined as:

$$\hat{Q}_X(\alpha) = X_{r:N} \text{ with } \frac{r - 1}{N} < \alpha \le \frac{r}{N}. \tag{6.19}$$

From (6.17) an asymptotic confidence interval for $Q_X(\alpha)$ with confidence level $1 - p$ can be derived:

$$\hat{Q}_X(\alpha) = \pm z_{1-\frac{p}{2}} \sqrt{\frac{\alpha(1 - \alpha)}{N f_X^2(Q_X(\alpha))}}, \tag{6.20}$$

where $z_{1-\frac{p}{2}} = \Phi^{-1}\left(1 - \frac{p}{2}\right)$, the inverse of the cumulative distribution function of the standard normal distribution.

Unfortunately this confidence interval is of little practical use as it requires knowledge of f_X. However, some insight can be gained about the accuracy of the quantile estimates. Let's suppose that F_X is the cumulative standard normal distribution. The plot of the 95% confidence bands for sample sizes ranging from 100 to 1000 for quantile levels of 1%, 3%, 5%, 10%, and 50% are displayed in Fig. 6.2. Obviously the accuracy of the estimates is positively

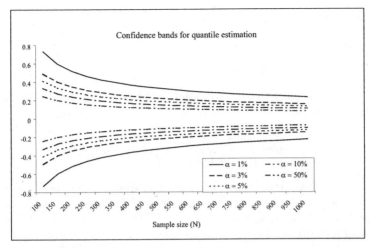

Figure 6.2. Confidence intervals of quantile estimate, $F_X = \Phi(0, 1)$

related to the sample size and depends on the value of α. Estimation of statistics located in one tail of the distribution suffers from high variability.

Exact confidence intervals for the population quantile based on order statistics and without assumption on F_X can be obtained as follows:

Search l and u so that:

$$P(X_{l:N} < q_\alpha < X_{u:N}) = 1 - p \text{ with } 0 < p < 1. \qquad (6.21)$$

Clearly:

$$P(X_{l:N} < q_\alpha < X_{u:N}) = P(X_{l:N} < q_\alpha) - P(X_{u:N} < q_\alpha). \qquad (6.22)$$

and with F_X strictly increasing:

$$X_{l:N} < q_\alpha \text{ iff } F_X(X_{u:N}) < F_X(q_\alpha). \qquad (6.23)$$

From the probability-integral theorem it is well-known that the distribution of the random variable $F_X()$ is uniform over the interval $[0, 1]$ (see 5.45 for a development). $F_X(X_{r:N})$ is therefore the rth order statistic from the uniform distribution. Moreover, $F_X(q_\alpha) = \alpha$ by definition.

Thus from (6.8):

$$P(X_{l:N} < q_\alpha) = P(F_X(X_{l:N})) < \alpha$$
$$= \int_0^\alpha \frac{N!}{(l-1)!(N-l)!} x^{l-1}(1-x)^{N-l} dx. \qquad (6.24)$$

For a given p, l and u have therefore to be chosen so that:

$$1 - p = \int_0^\alpha \frac{N!}{(l-1)!(N-l)!} x^{l-1}(1-x)^{N-l} dx$$
$$- \int_0^\alpha \frac{N!}{(u-1)!(N-u)!} x^{u-1}(1-x)^{N-u} dx. \tag{6.25}$$

After the integration by parts of the left hand-side expression of (6.25) the equality reduces to:

$$1 - p = \int_{n=l}^{u-1} \frac{N!}{(n)!(N-n)!} \alpha^n (1-\alpha)^{N-n}. \tag{6.26}$$

It should be noted that for arbitrary p, it is generally not possible to find integers l and u, so that (6.26) holds exactly. Let's consider the order statistics $X_{l:N}$ and $X_{u:N}$ with that $l = \alpha - 2\%$ and $u = \alpha + 2\%$. Figure 6.3 depicts the probability that q_α lies in the confidence interval bounded below by $X_{l:N}$ and above by $X_{u:N}$ for selected values of α and for $N \in [100, 1000]$.

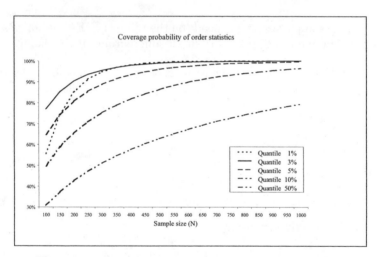

Figure 6.3. Confidence intervals based on order statistics

The coverage probability increases with the sample size. Clearly more data leads to a more accurate estimation of q_α. More interesting is the fact that for a given sample size the coverage probability increases with decreasing α, which seems at first glance to be little intuitive. In fact, in the interior the sampling distribution is very smooth with order statistics lying very close to each other. On the opposite, in the tails the interval between adjacent order statistics is much larger, which explains the high coverage probability. As a consequence of the lack of smoothness in the tails, the quantile estimates exhibit a rather erratic behaviour.

6.3 Quantile Estimators Based on Order Statistics

As shown above (see Sect. 6.2.3), $X_{r:N}$ with $r = [N\alpha] + 1$ is a consistent estimator of the αth-quantile of F_X. Moreover, the link between order statistics and the sample distribution F_N is immediate and the relationship between the sample distribution and the population distribution is given for instance by the *Glivenko-Cantelli theorem* (see appendix A.5):

$$P\left(\sup_{x \in \mathbb{R}} \mid F_N(x) - F_X(x) \mid \to 0\right) = 1. \tag{6.27}$$

The sample distribution converges uniformly with probability one to $F_X(x)$. A logic estimator of the population quantile is therefore $x_{r:N}$, the rth observation of the ordered sample of size N with $r = [N\alpha] + 1$. However, the derivation of this estimator makes use of asymptotical arguments. In practice N is finite and can be relatively small. Due to the discreteness of F_N, any value between $x_{r-1:N}$ and $x_{r:N}$ could act in fact as an estimator of the α-quantile. The literature proposes therefore several estimators based on the combination of the adjacent observations $x_{r-1:N}$ and $x_{r:n}$. The following estimators are reviewed in Parrish (1990) and Dielman et al. (1994):

1. $\left\{\begin{array}{l}\text{Lower empirical CDF value} \\ \text{where } r = [N\alpha] + 1\end{array}\right.$	E1	$= x_{r-1:n}$
2. $\left\{\begin{array}{l}\text{Upper empirical CDF value} \\ \text{where } r = [N\alpha] + 1\end{array}\right.$	E2	$= x_{r:n}$
3. $\left\{\begin{array}{l}\text{Weighted average at } X_{[N\alpha]:N} \\ \text{where } r = [N\alpha] + 1 \\ \text{and } g = [N\alpha] - r + 1\end{array}\right.$	E3	$= (1-g)x_{r-1:N} + gx_{r:N}$
4. $\left\{\begin{array}{l}\text{Observation numbered closest} \\ \text{to } N\alpha\end{array}\right.$	E4	$= \left\{\begin{array}{ll} x_{r-1:N} & \text{if } g < 0.5 \\ x_{r:N} & \text{if } g \geq 0.5 \end{array}\right.$
5. $\left\{\begin{array}{l}\text{Empirical CDF where } g = N\alpha - j \\ \text{and } j = [N\alpha]\end{array}\right.$	E5	$= \left\{\begin{array}{ll} x_{r-1:n} & \text{if } g = 0 \\ x_{r:n} & \text{if } g \geq 0 \end{array}\right.$
6. $\left\{\begin{array}{l}\text{Weighted average at } X_{[(N+1)\alpha]:N} \\ \text{where } k = [(N+1)\alpha] + 1 \\ \text{and } h = (N+1)\alpha - k + 1\end{array}\right.$	E6	$= (1-h)x_{k-1:N} + hx_{k:N}$
7. $\left\{\begin{array}{l}\text{Empirical CDF with averaging} \\ r, j \text{ like 3.}\end{array}\right.$	E7	$= \left\{\begin{array}{ll} \frac{(x_{r-1:N}+x_{r:N})}{2} & \text{if } g = 0 \\ x_{r:N} & \text{if } g \geq 0 \end{array}\right.$
8. $\left\{\begin{array}{l}\text{Weighted average at } X_{[N\alpha+0.5]:N} \\ \text{where } 0.5 \leq N\alpha \leq N - 0.5 \\ \text{where } i = [N\alpha + 0.5]\end{array}\right.$	E8	$= \begin{array}{l}(0.5 + i - N\alpha)x_{i:N} \\ + \\ (0.5 - i + N\alpha)x_{i+1:N}\end{array}$

If $N\alpha$ = any integer, $g = 0$ and therefore E1, E3, E4, and E5 are equivalent. Additionally, E7 = E8. E4 = E8 if $(N\alpha + 0.5)$ is an integer and $0.5 \leq N\alpha \leq N - 0.5$. Finally, for $(N + 1)\alpha$ = any integer, E2 is equivalent to E6.

6.3.1 Linear Combination of Several Order Statistics

For small samples, the estimators reviewed above suffer from the variability of the individuals or adjacent order statistics. A way to reduce this variability is to consider linear combinations of order statistics. These estimators based on weighted averages of the order statistics are referred to as *L estimators*. Clearly most weight is put on the order statistics $X_{r:N}$ for which $\frac{r}{N}$ is close to α. The estimators proposed by Harrell and Davis (1982) and Kaigh and Lachenbruch (1982) are introduced here.

The gamma and beta functions are defined as:

$$\Gamma(z) = \int_0^{+\infty} t^{z-1}e^{-t}dt, \tag{6.28}$$

$$\beta(z, w) = \frac{\Gamma(z)\Gamma(w)}{\Gamma(z + w)}. \tag{6.29}$$

When the argument z is an integer:

$$\Gamma(n + 1) = (n)!, \tag{6.30}$$

and therefore for z and w integers:

$$\beta(z + 1, w + 1) = \frac{(z)!(w)!}{(z + w + 1)!}. \tag{6.31}$$

Thus, (6.16) can been rewritten as:

$$E[X_{r:N}] = \frac{N!}{(r-1)!(N-r)!} \int_0^1 F_X^{-1}(u)u^{r-1}(1-u)^{N-r}du$$

$$= \frac{1}{\beta(r, N-r+1)} \int_0^1 F_X^{-1}(u)u^{r-1}(1-u)^{N-r}du. \tag{6.32}$$

Since $E\left[X_{[N\alpha]+1:N}\right]$ converges to $F_X^{-1}(\alpha)$, Harrell and Davis (1982) propose as a quantile estimator:

$$E9 = \frac{1}{\beta\left((N+1)\alpha, (N+1)(1-\alpha)\right)} \int_0^1 F_N^{-1}(u)u^{(N+1)\alpha-1}(1-u)^{(N+1)(1-\alpha)-1}du$$

As shown in (6.12) $F_N(t) = \frac{1}{N} \sum_{n=1}^{N} I_{X_n \in (-\infty, t])}$. E9 can therefore be reexpressed as:

$$\sum_{n=1}^{N} W_{N,n} X_{n:N}, \tag{6.33}$$

with:

$$
\begin{aligned}
W_{N,n} &= \frac{\int_{(n-1)/N}^{n/N} u^{(N+1)\alpha-1}(1-u)^{(N+1)(1-\alpha)-1} du}{\beta\left((N+1)\alpha, (N+1)(1-\alpha)\right)} \\
&= I_{n/N}\left(\alpha(N+1), (1-\alpha)(N+1)\right) \\
&\quad - I_{(n-1)/N}\left(\alpha(N+1), (1-\alpha)(N+1)\right),
\end{aligned}
\tag{6.34}
$$

where $I_x(a, b)$ denotes the incomplete beta function.

Let $Y_{1:k;N}, \cdots, Y_{k:k;N}$ be the ordered observations of a subsample of size k selected without replacement from the complete sample $X_1, \cdots, X_N$. From the combinatorics:

$$P(Y_{r:k;N} = X_{j:N}) = \frac{(j-1)!(N-j)!k!(N-k)!}{(r-1)!(j-r)!(k-r)!(N+r-j-k)!N!}, \tag{6.35}$$

where $r \leq j \leq r + N - k$.

Kaigh and Lachenbruch (1982) propose as a quantile estimator the subsample quantile averaged over all $\frac{N!}{k!(N-k)!}$ subsamples of size k:

$$\boxed{E10 = \sum_{j=r}^{r+N-k} \left(\frac{(j-1)!(N-j)!k!(N-k)!}{(r-1)!(j-r)!(k-r)!(N+r-j-k)!N!}\right) X_{j:N}}$$

where $r = [(k+1)\alpha]$.

6.4 Kernel-Based Estimators

The two L estimators introduced in the last section are smooth, reflecting the fact they are a weighted mean of the order statistics. It can be shown that several L estimators are asymptotically *kernel estimators* with Gaussian kernel (see Sheather and Jones (1991)). Kernel estimators are based on specific weight functions called kernel functions (see Fig. 6.4). As the L estimators, kernel estimators are smooth and therefore less variable than the simple estimators based on one or two order statistics.

It is worth noting that kernel estimation has been developed for the estimation of the unknown density function f_X from a sample of identical distributed observations. This technique is therefore equally adequate for the estimation of risk measures based on moments of the portfolio distribution such as TailVaR, the central moments and the lower partial moments for instance. This section follows in part the conceptual layout of Silverman (1986), Chap. 3. Further literature is Härdle (1991).

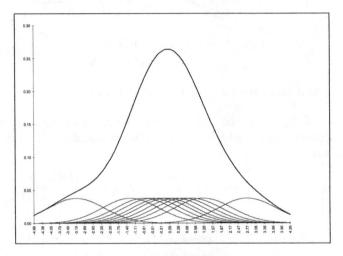

Figure 6.4. Density estimation with kernel smoothing

A kernel function K satisfies the condition:

$$\int_{-\infty}^{+\infty} K(x)dx = 1. \tag{6.36}$$

A kernel estimator $\hat{f}_X$ of density function f_X, based on observed or generated sample of size N, is defined by:

$$\hat{f}_X(x) = \frac{1}{Nh} \sum_{n=1}^{N} K\left(\frac{x - X_n}{h}\right), \tag{6.37}$$

where h is the window width, also called smoothing parameter or bandwidth. If h approaches zero, the representation of the data is very noisy. As h becomes large the density estimate becomes smoother and all detail is obscured (see Fig. 6.5).

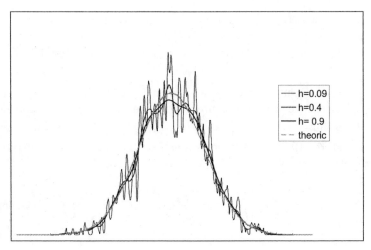

Figure 6.5. Density estimation with various bandwidths

Usually kernels have following properties (see Härdle (1991), p.45):

- Kernel functions are usually symmetric around zero and integrate to one.
- If the kernel is a density function, the kernel estimate is a density, too.
- The property of smoothness of kernels is inherited by $\hat{f}_X(x)$. If K is n times continuously differentiable, $\hat{f}_X(x)$ is also n times continuously differentiable.
- Kernel estimates do not depend on any choice of origins, as histograms do.

6.4.1 Accuracy of the Estimate Density

The accuracy of the density estimate $\hat{f}_X$ is a function of the kernel K and of the window width h. A traditional measure of the discrepancy between $\hat{f}_X$ and f_X is the *Mean Square Error* (MSE):

$$
\begin{aligned}
\text{MSE}_x(\hat{f}_X) &= E\left[\hat{f}_X(x) - f_X(x)\right]^2 \\
&= \left(E\left[\hat{f}_X(x)\right] - f_X(x)\right)^2 + \text{var}\left[\hat{f}_X(x)\right].
\end{aligned}
\tag{6.38}
$$

Therefore, MSE is the sum of the squared bias and the variance at x. In order to assess the global accuracy of an estimate for each x the squared error is integrated over the real line and the corresponding measure, called the *Mean Integrated Square Error* (MISE), is given by:

$$
\text{MISE}(\hat{f}_X) = E\left[\int \left(\hat{f}_X(x) - f_X(x)\right)^2 dx\right],
\tag{6.39}
$$

which is equivalent to:

$$\text{MISE}(\hat{f}_X) = \int \left(E\left[\hat{f}_X(x)\right] - f_X(x)\right)^2 dx + \int \text{var}\left[\hat{f}_X(x)\right] dx. \quad (6.40)$$

Minimising MISE reflects the desire to find a $\hat{f}_X$, which adequately estimates f_X everywhere. Alternative measures, e.g. L^1 measures like the expected integrated absolute error $\left(= E\left[\int \left|\left(\hat{f}_X(x) - f(x)\right)\right| dx\right]\right)$, yield similar results (see Jones et al. (1996b)). For kernel functions the mean and the variance of the estimate are given with:

$$E\left[\hat{f}_X(x)\right] = \frac{1}{N}\sum_{n=1}^{N} E\left[\frac{1}{h}K\left(\frac{x-X_n}{h}\right)\right] = \int \frac{1}{h}K\left(\frac{x-y}{h}\right) f_X(y)dy,$$
$$(6.41)$$

$$\text{var}\left[\hat{f}_X(x)\right] = \frac{1}{N}\int \frac{1}{h^2}K\left(\frac{x-y}{h}\right)^2 f_X(y)dy - \left\{\frac{1}{h}\int K\left(\frac{x-y}{h}\right) f_X(y)dy\right\}^2. $$
$$(6.42)$$

It is worth noting that the bias, contrary to the variance, does not depend on the sample size. The calculations of the MSE and MISE are often cumbersome and in practice approximations are used. For kernel functions with following properties (a justification for these properties is given by asymptotic arguments, see Silverman (1986), pp.70 ff):

$$\int K(t)dt = 1, \qquad \int tK(t)dt = 0 \qquad \text{and} \int t^2 K(t)dt = k_2 \neq 0, < \infty,$$
$$(6.43)$$

the bias is equal to:

$$\text{bias}\left[\hat{f}_X(x)\right] = E\left[\hat{f}_X(x)\right] - f_X(x). \quad (6.44)$$

Defining $y = x + ht$ and considering a Taylor expansion around x:

$$f_X(x+ht) = f_X(x) + htf'_X(x) + \frac{1}{2}h^2t^2 f''_X(x) + \text{ higher order terms,}$$
$$(6.45)$$

the bias can be approximated as follows:

$$\text{bias}\left[\hat{f}_X(x)\right] = \int \left[\frac{1}{h}K\left(\frac{x-y}{h}\right)f_X(y)\right]dy - f_X(x)$$

$$= \int \left[K(-t)f_X(x+ht)\right]dt - f_X(x)$$

$$= \int \left[K(t)f_X(x+ht) - f_X(x))\right]dt$$

$$\text{as } K \text{ integrates to 1 and } K(t) = K(-t)$$

$$= hf_X'(x)\int tK(t)dt + \frac{1}{2}h^2 f_X''(x)\int t^2 K(t)dt + \text{ higher order terms}$$

$$\approx \frac{1}{2}h^2 f_X''(x)k_2,$$

$$\text{as } \int tK(t)dt = 0 \text{ and } \int t^2 K(t)dt = k_2,$$

$$(6.46)$$

and therefore:

$$\int \text{bias}\left[\hat{f}_X(x)\right]^2 dx \approx \frac{1}{4}h^4 k_2 \int f_X''(x)^2 dx. \qquad (6.47)$$

A similar development gives as approximation for the variance:

$$\text{var}\left[\hat{f}_X(x)\right] \approx \frac{1}{Nh}f(x)\int K(t)^2 dt, \qquad (6.48)$$

and thus:

$$\int \text{var}\left[\hat{f}_X(x)\right]dx \approx \frac{1}{Nh}\int K(t)^2 dt,$$

$$\text{as } \int f_X(x)dx = 1. \qquad (6.49)$$

The fundamental problem of choosing an adequate h is now obvious: a very small h will tend to eliminate the bias but the variance will become large. The converse is naturally true. From above the *Approximate Mean Integrated Square Error* (AMISE) can be defined as:

$$\text{AMISE} = \frac{1}{4}h^4 k_2^2 \int f_X''(x)^2 dx + \frac{1}{Nh}\int K(t)^2 dt. \qquad (6.50)$$

The AMISE approximates the MISE asymptotically. Parzen (1962) shows that the bandwidth which minimises the AMISE (h_{opt}) is equal to:

$$h_{opt(\text{AMISE})} = \frac{1}{k_2^{2/5}N^{1/5}}\left\{\int K(t)^2 dt\right\}^{1/5}\left\{\int f_X''(x)^2 dx\right\}^{-1/5}. \qquad (6.51)$$

For convenience equations (6.50), respectively (6.51), may be rewritten:

$$\text{AMISE} = \frac{1}{4}h^4 k_2^2 R\left(f_X''\right) + \frac{R(K)}{Nh},\qquad(6.52)$$

respectively:

$$h_{opt(\text{AMISE})} = \left[\frac{R(K)}{NR(f_X'')k_2^2}\right]^{1/5},\qquad(6.53)$$

where $R(a) = \int a(x)^2\,dx$.

Hence, the ideal window width will converge to zero as the sample size increases, but at very slow rate. One shortcut of (6.53) is that h_{opt} depends on the unknown density to be estimated. Substituting the value h_{opt} in (6.52) gives:

$$\text{AMISE} = \frac{5}{4}\left[\frac{R(K)^4 R(f_X'')k_2^2}{N^4}\right]^{\frac{1}{5}}.\qquad(6.54)$$

An optimal kernel should therefore minimise $C(K) = k_2^{2/5}\{R(K)\}^{4/5}$. It can be shown that the kernel of the form $K(t) = \frac{3}{4\sqrt{5}}(1 - \frac{1}{5}t^2)$, the Epanechnikov kernel denoted by K_e, minimises $C(K)$ (see for instance Hodges and Lehmann (1956)). Comparing the efficiency of other symmetrical kernels K by using as efficiency ratio eff(K):

$$\text{eff}(K) = \left(\frac{C(K_e)}{C(K)}\right)^{5/4},\qquad(6.55)$$

it can be observed (see table 6.4.1 and Silverman (1986), p.43) that all efficiencies are very close to one. The choice of the kernel is therefore not very important with respect to the AMISE. The selection of the adequate bandwidth is much more crucial.

Kernel	K(t)		Efficiency	
Epanechnikov	$\frac{3}{4\sqrt{5}}(1-\frac{1}{5}t^2)$ 0	for$\|t\|<\sqrt{5}$ otherwise	1	
Biweight	$\frac{15}{16}(1-t^2)^2$ 0	for$\|t\|<1$ otherwise	$\left(\frac{3087}{3125}\right)^{1/2}$	≈ 0.9939
Triangular	1-$\|t\|<1$ 0	for$\|t\|<1$ otherwise	$\left(\frac{243}{250}\right)^{1/2}$	≈ 0.9859
Gaussian	$\frac{1}{\sqrt{2\pi}}e^{-(1/2)t^2}$		$\left(\frac{36\pi}{125}\right)^{1/2}$	≈ 0.9512
Rectangular	$\frac{1}{2}$ 0	for$\|t\|<1$ otherwise	$\left(\frac{108}{125}\right)^{1/2}$	≈ 0.9295

Table 6.2. Efficiency of selected kernels

6.4.2 Bandwidth Selection

Choosing an adequate h turns out to be the critical point in kernel density estimation. The visual choice of the smoothing parameter (the bandwidth) is a natural and very powerful method in the hands of experienced users. However, in the situation where many estimates are required, methods which provide an automatic choice for h are very useful. For financial institutions with a large number of continuously re-balanced portfolios an automatic bandwidth selector is a prerequisite. The literature about automatic bandwidth selection methods is growing fast and only a brief introduction on the topic will be given here. Jones et al. (1996b), Jones et al. (1996a), Sheather (1992), and Cao et al. (1994) provide good surveys of the recent development in this area. As mentioned above a bandwidth is optimal if it minimises the discrepancy between the true (and unknown) density f_X and the density estimator. Measures of discrepancy are the MISE, the AMISE or the integrated squared error (ISE). A definition of the ISE is given below. As shown above, the search of the optimal bandwidth is tricky as it depends on the unknown f_X. Following methods have been used for estimating h_{opt}

Rules of Thumb The simplest approach consists in replacing f_X by an arbitrary density function $\check{f}_X$. Thus, an approximate value $\int \check{f}_X''(x)^2 dx$ can be assigned to $\int f_X''(x)^2 dx$ in (6.52). If $\check{f}_X$ is assumed to be the normal density:

$$R\left(\check{f}_X''\right) = \frac{3}{8}\pi^{-1/2}\sigma^{-5}. \tag{6.56}$$

In case the standard deviation is chosen as a scale parameter, the optimal bandwidth for a Gaussian kernel is:

$$h_{RT1} = \left(\frac{4}{3}\right)^{1/5}\sigma N^{-1/5} = 1.06\,\sigma\,N^{-1/5}. \tag{6.57}$$

If the scale parameter chosen is the more robust interquartile range (IQR), i.e. the difference between the two sample quartiles, the optimal bandwidth for a Gaussian kernel becomes:

$$h_{RT2} = 0.79\,\text{IQR}\,N^{1/5}. \tag{6.58}$$

Least Squares Cross-Validation (LSCV) This method minimises the integrated squared error (ISE):

$$\begin{aligned}\text{ISE}(h) &= \int \left(\hat{f}_X(x,h) - f_X\right)^2 dx \\ &= R\left(\hat{f}_X(.;h)\right) - 2\int \hat{f}_X(x,h)f_X\,dx + R(f_X).\end{aligned} \tag{6.59}$$

Clearly, minimising the ISE is equivalent to minimising the first two terms of (6.59). If $\hat{f}_{X,n}(x,h)$ is the kernel estimate based on the $N-1$ data points other than X_n, then $\frac{2}{N}\sum_{n=1}^{N}\hat{f}_{X,n;h}(X_n)$ is an estimate of the second term (see Sheather (1992)). The bandwidth that minimises the last squares cross-validation function (LSCV) with:

$$\text{LCSV} = R\left(\hat{f}_X(.;h)\right) - \frac{2}{N}\sum_{n=1}^{N}\hat{f}_{X,n;h},(X_n) \qquad (6.60)$$

is referred to as the least squares cross-validation bandwidth and is denoted by h_{LSCV}. It has been a standard method of automatic smoothing but does not perform satisfactory: the method converges very slowly [relative rate $\frac{h_{LSCV}}{h_{opt}} = 1 + O_p(N^{-1/10})$], so that very large sample sizes are requested.

Biased Cross-Validation (BCV) This method attempts to minimise directly the AMISE. The requested approximation of the unknown $R(f_X'')$ is done by differentiating the kernel estimate $\hat{f}_X(x,h)$ twice, so that after squaring $\hat{f}_X(x,h)$ and integrating it:

$$R\left(\hat{f}_X''(.;h)\right) = \int \left[\hat{f}_X''(x,h)\right]^2 dx. \qquad (6.61)$$

This method minimises following expression:

$$\frac{R(K)}{nh} + \frac{1}{4}h^4 k_2^2 \left[R\left(\hat{f}_X''(.;h)\right) - \frac{R(K'')}{nh^5}\right]. \qquad (6.62)$$

Note that the expression $\frac{R(K'')}{nh^5}$ corrects the positive bias of the estimation as shown in Scott and Terrell (1987). The convergence of the BCV estimator is as slow as the LSCV estimate but the variability is much smaller.

Plug-in Methods Like the BCV, the plug-in methods attempt to minimise the AMISE by replacing $R(f_X'')$ by a consistent estimator. Unlike the BCV, the bandwidth $\check{h}$ used to estimate $R(f_X'')$ differs from the bandwidth h, which is intended to be adequate to estimate f_X. The reason is that bandwidths appropriate for curve estimation are quite different from those which are adequate for the estimation of $R(f'')$: for the derivation of the optimal $\check{h}$ see Hall and Marron (1987). A usual approach is to express $\check{h}$ as a function of h ($\check{h} = l(h)$) and then find the bandwidth which minimises the equation:

$$\left[\frac{R(K)}{Nk_2^2 R\left(\hat{f}_X''(.;l(h))\right)}\right]^{1/5}. \qquad (6.63)$$

For alternative derivations of $l(h)$, see Park and Marron (1990) and Sheather and Jones (1991). Plug-in estimators converge much faster than LSCV or BCV estimators. The estimator of Park/Marron converges at a relative rate of $1 + O_p(N^{-4/13})$ and the estimator of Sheather/Jones performs slightly better with a rate of $1 + O_p(N^{-5/14})$.

Root-N Methods A better approximation of the MISE can be obtained by extending its asymptotic expression. Hall, Sheather, Jones, and Marron (Hall et al. (1991)) show that:

$$\text{MISE}(h) = N(h) + o\left(\frac{1}{Nh} + o(h^6)\right), \tag{6.64}$$

where: $N(h) = \frac{R(K)}{Nh} + \frac{1}{4}h^4 k_2^2 R(f_X'') - \frac{1}{24}h^6 k_2 k_4 R(f_X''')$.

The extension adds an extra term of order h^6 to the AMISE. Hall, Sheather, Jones, and Marron show further than the minimiser of $N(h)$ is asymptotically equivalent to (see Sheather (1992)):

$$h = h_{\text{AMISE}} + J_2(h_{\text{AMISE}})^{3/5},$$

where $J_2 = \frac{k_4 R(f_X''')}{20 k_2 R(f_X'')}$.

The theoretical advantage of the root-N methods is their fast relative convergence rate of $1 + O_p(N^{-1/2})$, which corresponds to the best possible convergence rate (see Hall and Marron (1991)).

6.4.3 Quantile Estimation Based on the Kernel Density Method

From (6.37):

$$\hat{f}_X(x; h) = \frac{1}{Nh} \sum_{n=1}^{N} K\left(\frac{x - X_n}{h}\right),$$

it follows that the kernel-based estimation of the cumulative distribution function yields:

$$\hat{F}_X(x; h) = \frac{1}{N} \sum_{n=1}^{N} \int_{-\infty}^{x} K\left[\frac{x - X_n}{h} \quad d\left(\frac{x - X_n}{h}\right)\right]. \tag{6.65}$$

The kernel-smoothing version of the quantile function is therefore given by:

$$\hat{F}_X^{-1}(q; h) = \frac{1}{h} \int_0^1 K\left(\frac{q - y}{h}\right) \hat{F}_X^{-1}(y) dy. \tag{6.66}$$

Four estimators based on kernel density estimation with different automatic bandwidth choice will be considered. The first two base on the classical rules of thumb:

$$E11 = \hat{F}_X^{-1}(q) = \frac{1}{h_{RT1}} \int_0^1 K\left(\frac{q-y}{h_{RT1}}\right) \hat{F}_X^{-1}(y)dy$$

$$E12 = \hat{F}_X^{-1}(q) = \frac{1}{h_{RT2}} \int_0^1 K\left(\frac{q-y}{h_{RT2}}\right) \hat{F}_X^{-1}(y)dy$$

The third is obtained by using a root-N convergent bandwidth selector of Jones/Marron/Park (h_{JMP}) as described in Jones et al. (1991):

$$E13 = \hat{F}_X^{-1}(q) = \frac{1}{h_{JMP}} \int_0^1 K\left(\frac{q-y}{h_{JMP}}\right) \hat{F}_X^{-1}(y)dy$$

and finally the last estimator is based on the plug-in method of Jones/Sheather (h_{SJ}), which performs very well in Cao et al. (1994):

$$E14 = \hat{F}_X^{-1}(q) = \frac{1}{h_{SJ}} \int_0^1 K\left(\frac{q-y}{h_{SJ}}\right) \hat{F}_X^{-1}(y)dy$$

6.5 Comparison of the Quantile Estimators

In this section the accuracy of the different quantile estimators discussed above is assessed for the profit and loss distribution of Portfolio PF^{II}. The estimates of q_α, ($\alpha = 1\%, 3\%$, and 5%) based on a sample of size 10'000 are displayed in tables 6.3, 6.4 and 6.5. For this relatively large sample only the simple estimators based on one or two order statistics are considered. For smaller samples of size 1'000, respectively 200, the smooth estimators are also included. The results are displayed in tables 6.6, 6.7, and 6.8 respectively 6.9, 6.10, and 6.11. For the sample sizes considered several simple estimators are equivalent and therefore only the numbers for $E1$, $E2$, $E6$, and $E7$ are listed. For E14 h_{SJ} is estimated with the S-plus routine $width.SJ$ described in Venables and Ripley (1994) and for E13 h_{JMP} is computed with the XploRe routine $bwsel_{jmp}$.

The bias, the standard deviation, the MSE (mean squared error = squared bias + variance) and the MAD (mean absolute deviation = average of the absolute difference between the estimates and the benchmark) of the different estimates are evaluated. This is done by a second level Monte Carlo Simulation with sample size 100. The original idea of bootstrapping the original samples has been left out because of the smoothness problems related to nonparametric bootstrapping in tails of a distribution (see Efron and Tibshirani

(1993)). The MSE is an overall measure of the size of the estimation error and MAD is its robust equivalent. The benchmark is the empirical quantile of a sample with size 1'000'000.

VaR 5%				N=10'000	
Estimator	Mean	St. deviation	Bias	$\sqrt{MSE}$	MAD
Benchmark	6263.84				
E1	6267.72	99.10	3.88	99.17	81.05
E2	6263.16	98.74	-0.68	98.74	80.43
E6	6267.49	99.08	3.65	99.14	81.01
E7	6265.44	98.90	1.60	98.91	80.73

Table 6.3. Estimation of VaR 5% with sample size 10'000

VaR 3%				N=10'000	
Estimator	Mean	St. deviation	Bias	$\sqrt{MSE}$	MAD
Benchmark	7590.04				
E1	7596.35	138.43	6.31	138.57	108.18
E2	7586.30	139.31	-3.74	139.36	108.24
E6	7596.05	138.44	6.01	138.57	108.14
E7	7591.33	138.79	1.29	138.80	108.01

Table 6.4. Estimation of VaR 3% with sample size 10'000

VaR 1%				N=10'000	
Estimator	Mean	St. deviation	Bias	$\sqrt{MSE}$	MAD
Benchmark	10643.15				
E1	10658.85	247.86	15.70	248.36	192.35
E2	10631.00	247.66	-12.15	247.95	195.78
E6	10658.57	247.85	15.42	248.32	192.35
E7	10644.92	247.32	1.77	247.33	193.64

Table 6.5. Estimation of VaR 1% with sample size 10'000

As tables 6.3, 6.4, and 6.5 show, the estimation of the quantiles is relatively unproblematic when a large sample is at disposal. All estimators perform well and exhibit relatively small bias and standard deviation. As expected, the accuracy decreases with α. $E2$, the empirical quantile, is the best performer under the estimators based on one single order statistic. This is not surprising as $E2$ is the consistent estimator of q_α. However, considering the average of adjacent order statistics (E7) helps to reduce the bias, particularly for small α.

For the sample sizes 1000 and 200 the numbers show that smooth estimators have some gains in terms of variance reduction but most of them increase the bias. The Jones/Sheather estimator, E14, has a consistently good behaviour and perform at best with respect to the MAD and the MSE for almost all estimations. It has significantly less variance than the other estimators but exhibits a relative high bias, which suggests relative large bandwidths. On the opposite, the estimator of Kaigh/Lachenbruch, E10, exhibits a bad behaviour with a high bias as well as a high variance. The statistics for other estimators are not consistent enough to allow a precise ranking. For the intermediate sample size of 1000 the simple estimators, particularly the empirical quantile E2, have a small bias. For $N = 200$, the estimator based on the robust rule of thumb yields the best results with respect to the bias. The results show further that the achievement of good accuracy for very low quantiles is challenging. This is particularly true for small samples. At the 1% level for example, the standard deviation of the estimates amounts about 20% of the value of the benchmark. Thus, the width of a 95% confidence interval for q_α will amount about 80% of the value of q_α if we assume that the sampling distribution of q_α is approximatively Gaussian.

VaR 5%				N=10000	
Estimator	Mean	St. deviation	Bias	$\sqrt{MSE}$	MAD
Benchmark	6263.84				
$E1$	6325.11	346.42	61.27	351.80	280.25
$E2$	6269.90	340.10	6.06	340.15	274.86
$E6$	6322.35	345.93	58.51	350.84	279.61
$E7$	6297.50	342.32	33.66	343.97	275.63
$E9$	6317.28	332.24	53.44	336.51	269.44
$E10$	6345.02	334.33	81.18	344.04	274.92
$E11$	6367.58	325.00	103.74	341.15	272.66
$E12$	6328.02	329.26	64.18	335.45	267.55
$E13$	6327.49	329.66	63.65	335.74	268.02
$E14$	6428.71	308.51	164.87	349.80	276.62

Table 6.6. Estimation of VaR 5% with sample size 1000

	VaR 3%			N=1000	
Estimator	Mean	St. deviation	Bias	$\sqrt{MSE}$	MAD
Benchmark	7590.04				
E1	7627.69	479.03	37.65	480.50	402.73
E2	7550.25	470.61	-39.79	472.29	399.34
E6	7625.37	478.56	35.33	479.86	402.41
E7	7588.97	472.93	-1.07	472.93	399.87
E9	7644.72	461.53	54.68	464.75	389.68
E10	7692.08	467.48	102.04	478.49	396.47
E11	7670.70	450.48	80.66	457.64	381.77
E12	7629.80	457.48	39.76	459.21	386.83
E13	7630.03	457.15	39.99	458.90	387.77
E14	7728.20	429.73	138.16	451.40	367.84

Table 6.7. Estimation of VaR 3% with sample size 1000

	VaR 1%			N=1000	
Estimator	Mean	St. deviation	Bias	$\sqrt{MSE}$	MAD
Benchmark	10643.15				
E1	10936.14	918.57	292.99	964.17	795.10
E2	10629.11	847.65	-14.04	847.76	705.64
E6	10933.07	917.15	289.92	961.88	793.48
E7	10782.62	864.27	139.47	875.46	725.38
E9	10900.48	838.36	257.33	876.96	715.05
E10	11053.79	867.29	410.64	959.60	773.29
E11	10828.98	837.68	185.83	858.05	706.73
E12	10806.62	851.84	163.47	867.38	716.67
E13	10806.91	853.40	163.76	868.97	718.60
E14	10871.09	805.93	227.94	837.55	684.56

Table 6.8. Estimation of VaR 1% with sample size 1000

6.6 Summary

The accuracy of the estimation is clearly related to the size of the sample. Traditional quantile estimators based on one or two order statistics provide good results only for relative large samples. With smaller samples and therefore less information at disposal about the distribution, the use of smoothing techniques gives more insight into the data. Kernel based smoothers provide stable estimations in comparison with the simple estimators. However, the variability reduction is often linked with supplementary bias and the choice of the bandwidth influences significantly the accuracy of the estimation.

	VaR 5%			N=200	
Estimator	Mean	St. deviation	Bias	$\sqrt{MSE}$	MAD
Benchmark	6263.84				
$E1$	6314.82	789.36	50.98	791.01	615.85
$E2$	6077.12	742.11	-186.72	765.24	620.86
$E6$	6302.93	785.26	39.09	786.24	613.37
$E7$	6195.97	756.29	-67.87	759.33	599.76
$E9$	6332.66	718.62	68.82	721.91	558.27
$E10$	6473.97	744.38	210.13	773.47	599.66
$E11$	6354.51	698.45	90.67	704.31	543.53
$E12$	6288.66	717.87	24.82	718.30	560.36
$E13$	6282.18	720.98	18.34	721.22	563.59
$E14$	6585.21	624.39	321.37	702.24	541.40

Table 6.9. Estimation of VaR 5% with sample size 200

	VaR 3%			N=200	
Estimator	Mean	St. deviation	Bias	$\sqrt{MSE}$	MAD
Benchmark	7590.04				
$E1$	7690.16	1033.98	100.12	1038.81	817.14
$E2$	7260.17	942.13	-329.87	998.21	812.95
$E6$	7677.26	1028.62	87.22	1032.31	813.61
$E7$	7475.16	964.51	-114.88	971.33	786.43
$E9$	7739.59	958.94	149.55	970.53	761.65
$E10$	7999.54	1022.64	409.50	1101.58	871.06
$E11$	7624.01	936.16	33.97	936.78	748.49
$E12$	7557.65	951.14	-32.39	951.70	767.10
$E13$	7552.43	955.10	-37.61	955.84	770.99
$E14$	7838.35	858.29	248.31	893.48	698.89

Table 6.10. Estimation of VaR 3% with sample size 200

	VaR 1%			N=200	
Estimator	Mean	St. deviation	Bias	$\sqrt{MSE}$	MAD
Benchmark	10643.15				
$E1$	11286.47	1992.15	643.32	2093.45	1647.02
$E2$	9874.06	1678.99	-769.09	1846.76	1536.64
$E6$	11272.34	1984.86	629.19	2082.20	1639.35
$E7$	10580.26	1718.08	-62.89	1719.23	1391.24
$E9$	11299.96	1953.05	656.81	2060.53	1584.87
$E10$	12353.32	2631.29	1710.17	3138.21	2305.89
$E11$	10740.15	1799.90	97.00	1802.51	1417.83
$E12$	10627.96	1727.46	-15.19	1727.53	1400.16
$E13$	10679.55	1725.24	36.40	1725.63	1396.28
$E14$	10846.14	1605.09	202.99	1617.87	1285.98

Table 6.11. Estimation of VaR 1% with sample size 200

7. Conclusion and Outlook

7.1 Summary

Risk measurement represents the first step in the development of successful risk management: one cannot manage what one cannot measure. Clearly, no quantitative tool can replace the expertise and judgement of professionals. Risk measures, however, are essential decision supports and allow to synthesise the huge quantity of information available. A sound use of risk measures begins with a profound knowledge of the underlying assumptions, characteristics, and limitations. In this work the source of uncertainty, the risk factors, has been modelled as empirical quantities which represent measurable properties of the states of the world. As a result, the uncertainty faced by decision makers have been appropriately represented in probabilistic terms and the validity of the risk measures presented is linked to the probabilistic measurement of future events.

Chapter 2 reviews the most widely used risk measures derived from the portfolio distribution. It distinguishes between risk measures for investment and risk measures for capital requirement. Among the first class of risk measures the Lower Partial Moment (LPM) approach has a significant advantage over the generally used variance of making no assumption about the distribution of the portfolio. This is of particular importance for portfolios with large option content since their distribution may take any arbitrary form. In the class of risk measures for capital requirement TailVaR is especially attractive since it meets the subadditivity property. TailVaR measures the expected loss of a portfolio, given the loss exceeds VaR and can therefore be seen as a coherent extension of Value-at-Risk. It is interesting to note that LPM and TailVaR are closely related. Both LPM and TailVaR focus on the left tail of the portfolio distribution. As a result, a partial representation of the portfolio distribution, such as central moments, is not sufficient to describe the uncertainty of the model. Therefore, an adequate estimation of risk measures requires a careful modelling of the tails of the risk factor distributions. This is of particular importance since the daily log-returns of the risk factors exhibit excess kurtosis.

As discussed in *chapter 3*, models which accommodate leptokurtosis allow for changing levels of the conditional volatility. Among the models which have been discussed the Student-t has been implemented because of its theoretical and practical properties. Firstly, the model is able to capture the fat tail properties of short horizon returns. Secondly, the Student-t distribution converges to a normal one with increasing time horizon, which is supported by the empirical evidence. Thirdly, the model is very parsimonious, i.e. it requires the estimation of a small set of parameters, even for high dimensional problems. This latter property is particularly important for large portfolios which are frequently re-balanced. The empirical analysis of section 3.5 has shown that the Student-t model provides an excellent fit to the data and has significantly more descriptive power than the widely used generic Gaussian model. Note, however, that the Student-t model can not account for the volatility clustering, i.e. the empirical observed fact that large returns tend to be followed by large ones and small returns by small ones. Capturing volatility clustering is of particular interest for the volatility prediction and therefore for the pricing of contingent claims, such like options, where the volatility is a key input parameter. The use of a GARCH or a stochastic volatility model with Student-t distributed innovations would allow to take account of both the excess kurtosis and the correlation of the squared log-returns. However, for the approximation of risk measures the issue of leptokurtosis is particularly worrying since it implies that large losses will occur more frequently than predicted by the normal model. Estimates of risk measures based on the normal distribution of the log-returns may therefore be misleading. As shown in section 3.5.4 the alternative modelling of the risk factor distribution may have a severe impact on the portfolio distribution and therefore on the estimation of risk measures. The assumption of joint normality of the log-returns tends to underestimate the risk when the measure considered focuses on events occurring with small probability. The normality assumption is particularly dubious if the holding period considered is short or if the portfolio of interest is exposed to a small set of risk factors.

The review of the pricing of financial assets in *chapter 4* has put most emphasis on options. The evaluation of options is challenging since they exhibit a non-linear exposure to the risk factors, and many option pricing equations do not have a closed form solution. As a result, the full valuation of large portfolios with significant option content is computationally demanding. This is particularly important since the scenario-based methods used for the approximation of the portfolio distribution require the valuation of the portfolio for a whole set of scenarios. A standard procedure allowing to mitigate the computational burden consists in replacing the value function by a global Taylor approximation of first or second order. However, as shown in section 4.5.1, such approximations do not always converge to the true value, especially not if the horizon considered is large and if the scenarios considered are far away from the supporting point of the approximation. A significant improvement is achieved by the use of a piecewise quadratic approximation of the pay-off profile.

Chapter 5 deals with several approaches to approximate the portfolio distribution. The Delta approximation is clearly inaccurate for portfolios with option content, as the first order Taylor approximation can not cope with the non-linearity of the pay-off profiles. The Delta-Gamma approximation, based on global quadratic approximation, represents an improvement since it is able to capture some non-linearity of the price functions. However, the approximation worsens dramatically when the holding period considered is larger than a few days, that is when the potential market moves may increase (see, for instance, table 5.15). Moreover, the Delta-Gamma approximation assumes joint normality of the risk factor distribution, what is usual for the analytic approximations. As a result, the method cannot accommodate the excess kurtosis of the log-return distribution and is therefore inadequate for portfolios exposed to a small number of risk factors. Hence, the use of the Delta-Gamma approximation should be restricted to well-diversified and liquid trading portfolios for which the joint normality of the risk factors seem plausible and a short holding period is appropriate. If first condition holds, the issue of leptokurtosis will be not critical. The second condition ensures that the dispersion of the risk factor outcomes remains low.

The limitations of the analytic approximations may be relaxed in the light of scenarios-based methods which pose no assumption on the distribution information. The (crude) Monte Carlo simulation takes fully account of the whole non-linearity of the pay-off profiles, since the method is based on the full valuation of the portfolio for each generated scenario. Moreover, the accuracy of the method is independent of the dimension of risk factor space. Finally, the implementation of the Monte Carlo scheme is straightforward and does not involve much analytical skill. The major drawback of the method lies in its slow convergence, i.e. in the large number - several thousands - of evaluations required to achieve a good level of accuracy. For this reason, the Monte Carlo simulation is of little use for large portfolios, particularly not for portfolios whose pay-off profile is not given in an analytical form. On the other hand, the Monte Carlo simulation is adequate for small portfolios with highly non-linear profiles which are exposed to a large number of risk factors.

The BDPQA combines the flexibility of the Monte Carlo simulation with a reasonable numerical effort. Like the Monte Carlo simulation, the BDPQA poses no assumption on the risk factor distribution. The method combines full evaluation of the risk profile for a distinguished set of market scenarios with piecewise quadratic approximation of the value function. The supporting points of the piecewise quadratic approximation are given by the set of generalised barycenters obtained by refinement of the risk factor space. The number of generalised barycenters considered determines the number of full valuations and hence the computational burden associated with the BD-PQA. The refinement process is sequential and allows therefore for permanent control over the numerical effort. The result of section 5.5 has shown that the BDPQA yields accurate estimates after a small number of refinements. Hence, the piecewise quadratic approximation scheme considerably mitigate the numerical effort associated with the valuations of the portfolio for each scenario. Moreover, the results of the BDPQA consistently outperform those of the Delta-Gamma approximation for the non-linear portfolios considered. Finally, the accuracy of the BDPQA, unlike the Delta-Gamma approximation, is robust with respect to the length of the holding period considered. The issue of adequate holding period is discussed in sections. 5.5.1 and 7.2. The BDPQA is therefore particularly appropriate for large portfolios with option content, whose instruments are not necessarily liquid.

The output of both the BDPQA and the Monte Carlo simulation is a sample of size N whose empirical distribution is a proxy for the true portfolio distribution. For the Monte Carlo simulation the numerical effort is directly related to the size of the sample. *Chapter 6* investigates the impact of the sample size on the accuracy of the Value-at-Risk estimates. Sophisticated estimators, based on averaging methods such as linear combinations of order statistics or kernel smoothers, have been introduced. The main motivation for the use of these averaging methods has been the attempt to obtain accurate estimates from a small sample and therefore to reduce the number of portfolio valuations. The results show that the achievement of high accuracy in the measurement of tail probabilities is not straightforward when the sample at disposal is small and hence the information about the unknown portfolio distribution is limited. Estimators based on linear combinations of order statistics or kernel smoothers exhibit less variability than the classical ones but suffer from a larger bias. Clearly, no sophisticated smoothing technique can expand the information set.

Consequently, accuracy dictates large samples. On the other hand, the number of full evaluations one may afford is limited. These two features confirm the usefulness of the BDPQA, which is able to combine a large sample size with a limited number of full valuations.

7.2 The Issue of Credit Risk

Credit risk arises from changes in value due to credit events, such as defaults or changes in the credit quality of a counterparty. Only the issue of default risk is considered in this section. Default risk depends principally on the following components:

- the credit exposure, which is the amount lost if a counterparty must default, i.e. the replacement value of the position,
- default rate,
- the recovery rate, that is the proportion of the position that is likely to be recovered if the counterparty defaults.

The assessment of default risk requires the modelling of the joint probability distribution combining the credit exposure, default rate and recovery rate distributions. Note that the distributions of default and recovery rates are generally more difficult to model than the risk factor distribution since the data at disposal is sparse. The credit exposure at time t is given by:

$$\max(RV(t), 0) \tag{7.1}$$

Where $RV(t)$ is the replacement value of the position. Clearly, the credit exposure corresponds to the market value, if positive, of the position at time t. For traditional assets, such as loans and bonds, the exposure is close to the nominal value for every t. In contrast, the exposure of derivatives may vary considerably over the life of the instrument. Plain vanilla interest swaps, for instance, do not exhibit any credit exposure at initiation since each party has entered a zero-sum contract, i.e. the market value of the swap is zero. Over time the change in the market conditions affects the present value of the cash flows and the market value becomes positive or negative creating a credit exposure of one counterparty to the other. The potential exposure is the result of two offsetting factors: the amortisation effect and the diffusion effect. Over time, payments are made and the exposure goes down, hence the contract amortises gradually. On the other hand, as time elapses the dispersion of the interest rates increases which leads to an increase of the potential exposure. Due to the combination of both effects, the swap exposure is traditionally zero at origination, peaks somewhere in the mid-term, and declines towards zero as the contract approaches maturity. As a result, the analysis of default risk requires the modelling of the exposure over the whole life of the position. Hence, the holding period can yield several years.

7.3 Outlook

Several issues are subject to further research. As stated above, the use of a GARCH or a stochastic volatility model with Student-t innovations for the description of the risk factor dynamics would be an improvement on the unconditional Student-t model. The choice of the same underlying process for option pricing would ensure consistent treatment of the probability information over the whole risk measurement process. Further, the improvement of the BDPQA may be achieved through the development of more sophisticated refinement schemes. In particular the use of the geometrical information which is available during the discretisation of the risk factor space appears to be of major importance for accelerating the convergence behaviour. Finally, as briefly discussed in the last section, market and credit risks are closely related. The simultaneous modelling of both sources of financial risks is a further step towards an integrated risk management.

A. Probability and Statistics

This appendix summarises the probabilistic concepts and results used in this book for the modelling of uncertainty. The appendix follows in part the conceptual layouts of Brémaud (1994), Durrett (1996), Schervish (1995), and Bingham and Kiesel (1998). It is also referred to the quotations in the main text.

A.1 Probabilistic Modelling

The main objective of probability theory is to associate a likelihood to events arising from a random experiment. The result of any random experiment is called an *outcome* and is denoted by ω. Every outcome, also called *elementary event, ω* is distinguishable and indecomposable. Every collection of outcomes is called an *event*. Under uncertainty, the outcome of an experiment is not known in advance. However, the collection of all possible outcomes (possible "states of the world"), referred to as the *sample space* or *sure event Ω*, is assumed to be known a priori. Ω may be finite as well as countably or uncountably infinite and therefore a probability is not necessarily defined on all elements of Ω but rather on a collection $\mathcal{F}$ of subsets of Ω which satisfies a certain structure.

Formally, an *algebra* (a *field*) $\mathcal{F}_0$ is a collection of subsets of Ω with following properties:

i. $\Omega \in \mathcal{F}_0$,
ii. $A \in \mathcal{F}_0 \Rightarrow A^c = \Omega \setminus A \in \mathcal{F}_0$,   ($\mathcal{F}_0$ is closed under complementation)
iii. $A_1, A_2, \in \mathcal{F}_0 \Rightarrow A_1 \cup A_2 \in \mathcal{F}_0$,   ($\mathcal{F}_0$ is closed under finite union)

An algebra $\mathcal{F}$ is called a *σ-algebra* (*σ-field, tribe*) on Ω if, in addition:

iv. $\bigcup_{i \in \mathbb{N}} A_i \in \mathcal{F}$.   ($\mathcal{F}$ is closed under countable union)

Hence σ in this context means countably infinite. $\mathcal{F}$ contains all subsets of Ω for which a probability is associated. $\mathcal{F}$ represents the information or knowledge available about the possible states of the world. If Ω is finite, property (iv) is superfluous since Ω has finitely many subsets. The largest

σ-algebra is the *power set* with cardinality 2^{Ω} which is made up of all subsets of Ω. The smallest algebra is the *trivial algebra* $\{\varnothing, \Omega\}$, where $\varnothing = \Omega^c$ is called the *impossible event*. Note that for a given collection $\mathcal{A}$ of subsets of Ω, the smallest σ-algebra containing $\mathcal{A}$, i.e. the intersection of σ-algebras including $\mathcal{A}$, is called the *σ-algebra generated by $\mathcal{A}$*. In particular, the σ-algebra generated by the half-open intervals of $\mathbb{R}^M$ is called the *Borel σ-algebra* and denoted by $\mathcal{B}^M$. Its elements are the Borel sets $\mathcal{B}^M$ of $\mathbb{R}^M$.

The pair $(\Omega, \mathcal{F})$ is called a *measurable space*, i.e. a space on which a measure can be put. A *measure* on $(\Omega, \mathcal{F})$ is the mapping $\mu : \mathcal{F} \to \mathbb{R}$ which satisfies:

  i. $\mu(A) \geq 0 \quad \forall A \in \mathcal{F}$,
 ii. $\mu(\varnothing) = 0$,
iii. $\mu\left(\bigcup_{i \in \mathbb{N}} A_i\right) = \sum_{i \in \mathbb{N}} \mu(A_i)$.

The measure λ on the Borel σ-algebra with $\lambda((a, b]) = b - a$ for any $b > a \in \mathbb{R}$ exists and is called the *Lebesgue measure* on the real line. There is a unique (product) measure μ defined on the σ-algebra generated by the set of all Cartesian products $A_1 \times \cdots \times A_M$ such that:

$$\mu(A_1 \times \cdots \times A_M) = \mu_1(A_1)\mu_2(A_2)\cdots\mu_M(A_M).$$

The product measure on $\mathbb{R}^M$ is denoted by λ^M and called the Lebesgue measure on $\mathbb{R}^M$. For subsets of Ω with "familiar shape" the Lebesgue measure on $\mathbb{R}$, $\mathbb{R}^2$, respectively $\mathbb{R}^3$, coincides with the usual length, area, respectively volume. The triple $(\Omega, \mathcal{F}, \mu)$ is referred to as *measure space*. A measure P on $(\Omega, \mathcal{F})$ for which:

$$P(\Omega) = 1 \tag{A.1}$$

holds is called a *probability*. Hence the domain of P are the subsets of $\mathcal{F}$ and its range is the interval $[0, 1]$. The triple $(\Omega, \mathcal{F}, P)$ is termed a *probability space*. Note that following properties:

  i. $P(\varnothing) = 0$, $P(\Omega) = 1$,
 ii. $P(A) \geq 0$,
iii. $P\left(\sum_{i \in \mathbb{N}} A_i\right) = \sum_{i \in \mathbb{N}} P(A_i)$,
 iv. $A_2 \subseteq A_1$ and $P(A_1) = 0$, then $P(A_2) = 0$,
     (immediate since $P(A_2) = P(A_1) - P(A_1 \setminus A_2)$)

are the well known *Kolmogorov axioms of probability* (see Kolmogorov (1933)).

Two probability measures P and Q defined on the same σ-algebra $\mathcal{F}$ are said to be *equivalent* if they share the same null sets, i.e.:

$$P(A) = 0 \Leftrightarrow Q(A) = 0, \quad \forall A \in \mathcal{F}. \tag{A.2}$$

Hence, P and Q may associate different probabilities to events but agree as to which events are negligible. If every P-negligible event is Q-negligible but the converse is not necessarily true, i.e.:

$$P(A) = 0 \Rightarrow Q(A) = 0, \quad \forall A \in \mathcal{F}, \tag{A.3}$$

Q is said to be *absolutely continuous with respect to P*, which is written:

$$Q << P. \tag{A.4}$$

The *Radon-Nikodým* theorem states that Q is absolutely continuous with respect to P iff there exists a $\mathcal{F}$-measurable function f such that:

$$Q(A) = \int_A dQ = \int_A f dP, \quad \forall A \in \mathcal{F} \tag{A.5}$$

The function f (the *Radon-Nikodým derivative* of Q with respect to P) is often symbolically denoted by dQ/dP and hence:

$$\int_A dQ = \int_A \frac{dQ}{dP} dP. \tag{A.6}$$

If P and Q are equivalent then both dQ/dP and dP/dQ exist and:

$$\frac{dQ}{dP} = \left[\frac{dP}{dQ}\right]^{-1}. \tag{A.7}$$

A.2 Random Variable

Let's consider two measurable spaces $(\Omega, \mathcal{F})$ and $(\Omega', \mathcal{F}')$. A *measurable map* (or *random quantity*) is a function $X : \Omega \to \Omega'$ which satisfies:

$$X^{-1}(B) = \{\omega \in \Omega : X(\omega) \in B\} \in \mathcal{F} \quad \forall B \in \mathcal{F}'. \tag{A.8}$$

The smallest σ-algebra containing all sets:

$$\{\omega \in \Omega : X(\omega) \in B\}$$

is called the σ-algebra generated by X and noted $\sigma(X)$. $\sigma(X)$ represents the information contained in X. Note that X is $\mathcal{F}$-measurable if and only if $\sigma(X) \subseteq \mathcal{F}$.

A measurable map X induces a measure P_X on $(\Omega', \mathcal{F}')$ from P:

$$P_X(B) = P\left(X^{-1}(B)\right), \quad \forall B \in \mathcal{F}'. \tag{A.9}$$

P_X is referred to as the *distribution* of X. A measurable map assuming values in $\Omega' = \mathbb{R}^M$, or more precisely in $(\mathbb{R}^M, \mathcal{B}^M)$, is called a *random variable*.

If $M > 1$, a random variable is also called a *random vector*. If X and Y induce the same distribution on $(\mathbb{R}^M, \mathcal{B}^M)$, X and Y are said to be *equal in distribution*, which is denoted by $X \overset{d}{=} Y$.

A random variable X is said to be *simple* if it assumes only finitely many distinct values. A random variable X is called *discrete* if there exists a countable set $B \subseteq \Omega'$ such that $P_X(B) = 1$ and there exists a function $p_X(x)$, such that:

- $p_X(x) \geq 0$,
- $\sum_x p_X(x) = 1$,
- for every interval $[a, b]$:

$$P(a \leq X \leq b) = \sum_{x=a}^{b} p_X(x).$$

$p_X(x)$ is called the *frequency* or *probability mass* function of X.

A random variable is said to be *continuous* if there exists a function $f_X(x)$, such that:

- $f_X(x) \geq 0$,
- $f_X(x)$ exhibits at most a finite number of discontinuities in every finite interval of $\mathbb{R}$,
- $\int_{-\infty}^{\infty} f_X(x)dx = 1$,
- for every interval $[a, b]$:

$$P_X(a \leq X \leq b) = \int_a^b f_X(x)dx.$$

$f_X(x)$ is called the (probability) *density* function of X.

A.2.1 Distribution Function

A probability P_X defines a unique *cumulative distribution* function (CDF) $F_X(x)$ with:

$$F_X(x) = P_X(\{\omega \in \Omega : X(\omega) \leq x\}), \tag{A.10}$$

or as shorthand:

$$P_X(X \leq x). \tag{A.11}$$

The converse is also true. The properties of F_X are:

- F_X is nondecreasing,

- $\lim_{x\to\infty} F_X(x) = 1$, $\lim_{x\to-\infty} F_X(x) = 0$,
- F_X is right continuous, i.e. $\lim_{y\downarrow x} F_X(y) = F_X(x)$,
- If $F_X(x_-) = \lim_{y\uparrow x} F_X(y)$ then $F_X(x_-) = P(X < x)$,
- $P_X(X = x) = F_X(x) - F_X(x_-)$.

Obviously, the CDF of a discrete random variable X is given with:

$$F_X(x) = \sum_{y\leq x} p_X(y),$$ (A.12)

and its equivalent for random variable of a continuous type is:

$$F_X(x) = \int_{-\infty}^{x} f_X(y)dy.$$ (A.13)

If P_X is a probability measure on $(\mathbb{R}, \mathcal{B})$, the corresponding CDF $F_X(x)$ is equals to $P_X(X \in (-\infty, x])$. If P_X is a measure on $(\mathbb{R}^M, \mathcal{B}^M)$, the joint CDF of $X = (X_1, \cdots, X_M)$ is given with:

$$F_X(x_1, \cdots, x_M) = P_X(X_1 \in (-\infty, x_1], \cdots, X_M \in (-\infty, x_M]),$$ (A.14)

the function $F_{X_1}(x)$ with:

$$F_{X_1} = F_X(x_1, \infty, \cdots, \infty)$$ (A.15)

is called the *marginal distribution* function of X_1. The corresponding density function $f_{X_1}(x)$, if it exists, is called the marginal density of X_1.

A.2.2 Moments

The *expectation* (expected value) of a random variable X on $(\Omega, \mathcal{F}, P)$, if it exists, is denoted by $E[X]$ and is equal to:

$$\int_{\Omega} X(\omega)dP(\omega) = \int_{-\infty}^{\infty} x dF_X(x).$$ (A.16)

Hence the expectation of a random variable represents a weighted average over the probability space Ω and gives an indication of the central value of the density function when it exists. The expected value is therefore referred to as a *location parameter*. The expectation of a non-negative random variable is given with:

$$\lim_{n\to\infty} \sum_{k=0}^{n2^n-1} \frac{k}{2^n} P\left(\frac{k}{2^n} \leq X < \frac{k+1}{2^n}\right) + nP(X \geq n).$$ (A.17)

For a random variable X, which is not non-negative, define $x^+ = \max\{x, 0\}$ be the positive part and let $x^- = \max\{-x, 0\}$ be the negative part of x.

$E[X] = E[X^+] - E[X^-]$ if not both $E[X^+]$ and $E[X^-]$ are infinite. If $E[|X|] < \infty$, X is said to be integrable and then $E[X]$ is a finite number. Note that the expectation of X exists if:

$$\int_\Omega |X| \, dP < \infty, \tag{A.18}$$

and X is said to be *integrable*. The family of integrable random variables is denoted by L^1 (see also Sect. A.4).

If k is a positive integer then $E[X^k]$, if it exists, is called the kth *moment* of X. The first moment $E[X]$ is also called the mean and denoted by μ. The kth *central moment* of X, if it exists, is equal to $E[x - \mu]^k$. If $E[X^2] < \infty$ the variance of X is defined to be $\text{var}[X] = E[X - \mu]^2 = E[X^2] - 2\mu E[X] + \mu^2 = E[X^2] - \mu^2$. The square root of the variance is called the *standard deviation* $\text{std}[X]$. The standard deviation is an indication of how dispersed the distribution is about μ and is therefore a *scale parameter*. $E[X - \mu]^3$ provides information about the asymmetry of the distribution about its mean, which is also called the *skewness*. The fourth central moment gives insight into the relative weight of the centre of the distribution, the *kurtosis*. The classical measures of skewness γ and of kurtosis κ are the normalised third, respectively fourth, moment of the distribution:

$$\gamma = E\left[\frac{(X - \mu)^3}{\sigma^3}\right],$$
$$\kappa = E\left[\frac{(X - \mu)^4}{\sigma^4}\right]. \tag{A.19}$$

The *moment generating function* $\Psi(s)$ (or $M(s)$) of a random variable X is given with:

$$\Psi_X(s) = \int_{-\infty}^{\infty} e^{s^\top x} dF_X(x) = E\left[e^{s^\top X}\right], \tag{A.20}$$

the *characteristic function* $\phi(t)$ of X is defined as:

$$\phi_X(s) = \int_{-\infty}^{\infty} e^{is^\top x} dF_X(x) = E\left[e^{is^\top X}\right], \tag{A.21}$$

where $i = \sqrt{-1}$. Note that (A.20) represents the bilateral Laplace transform of the distribution F_X and the function needs not to exist for any t. The characteristic function of X is the Fourier-Stieltjes transform of F_X and always exists. If the density function of X exists, the right-hand side of (A.21) is the ordinary Fourier transform of F_X. The moment generating function, if it exists, and the characteristic function of X are unique and completely determine the distribution of the random variable. Hence, random variables with the same characteristic function are equal in distribution.

In the event of independence among the random variables (see Sect. A.2.3), the characteristic function of a sum of random variables is equal to the product of the characteristic functions. For instance, if X and Y are independent:

$$\phi_{X+Y}(s) = \left[e^{is^\top(X+Y)}\right] = \left[e^{is^\top X}\right]\left[e^{is^\top Y}\right] = \phi_X(s)\phi_Y(s). \qquad \text{(A.22)}$$

A.2.3 Independence and Correlation

The events $A_1, \cdots, A_N$ are called *independent* (for a probability measure P) if for every $J \subset \{1, \cdots, N\}$:

$$P\left(\bigcap_{j \in J} A_j\right) = \prod_{j \in J} P(A_j). \qquad \text{(A.23)}$$

Random variables $X_1, \cdots, X_N$ are (jointly) independent if:

$$P\left(\bigcap_{n=1}^N \{X_n \in B_n\}\right) = \prod_{n=1}^N P(\{X_n \in B_n\}). \qquad \text{(A.24)}$$

A necessary and sufficient condition of independence is:

$$P\left(\bigcap_{n=1}^N \{X_n \leq x_n\}\right) = \prod_{n=1}^N P(\{X_n \leq x_n\}), \qquad \text{(A.25)}$$

for all $x_1, \cdots, x_N \in (-\infty, \infty]$. If the random variables $X_1, \cdots, X_N$ are independent, the set $\{X_1, \cdots, X_N\}$ is called independent and every subset of it is independent. Random variables X_i ($i = 1, \cdots, N$) are called *pairwise independent* if, and only if, for all $i \neq j$, X_i and X_j are independent. Clearly, if the set $\{X_1, \cdots, X_N\}$ is independent the random variables X_i are pairwise independent. The converse is not necessarily true.

If $X_1, \cdots, X_N$ are independent with $E[|X_n|] < \infty$, then:

$$E\left[\prod_{n=1}^N X_n\right] = \prod_{n=1}^N E[X_n]. \qquad \text{(A.26)}$$

If X and Y are independent, then $g_1[X_1]$ and $g_2[X_2]$, where g_1 and g_2 are measurable functions, are independent as well. The *covariance* of two random variables X_i and X_j, written $\text{cov}[X_i, X_j]$, is defined by:

$$\text{cov}\,[X_i, X_j] = E\left[(X_i - E[X_i])(X_j - E[X_j])\right] = E[X_i X_j] - E[X]E[Y]. \qquad \text{(A.27)}$$

A family of random variables $X_i, i \in I$, with $E[X_i^2] < \infty$ is said to be *uncorrelated* if:

$$E[X_i X_j] = E[X_i]E[X_j], \text{ whenever } i \neq j \qquad (A.28)$$

holds. The *correlation coefficient* of X_i and X_j, $\lambda[X_i, X_j]$, is given with:

$$\lambda[X_i, X_j] = \frac{\text{cov } [X_i, X_j]}{\text{var } [X_i] \text{ var } [X_j]}. \qquad (A.29)$$

From equations (A.26), (A.27), and (A.28), it is easy to verify that independent variables are uncorrelated. The converse is not necessarily true.

A.2.4 Conditional Probability and Expectation

Let consider $A_1, A_2 \in \mathcal{F}$ such that $P(A_2) \neq 0$. The *conditional probability* of A_1, given A_2, is denoted $P(A_1 \mid A_2)$ and defined by:

$$P(A_1 \mid A_2) = \frac{P(A_1 \cap A_2)}{P(A_2)}. \qquad (A.30)$$

Clearly, if the events A_1 and A_2 are independent:

$$P(A_1 \mid A_2) = P(A_1). \qquad (A.31)$$

For any integrable random variable on a given space $(\Omega, \mathcal{F}, P)$ and any event $A \in \mathcal{F}$ with $P(A) > 0$ the conditional expectation of X, given A, is defined by:

$$E[X \mid A] = \frac{1}{P(A)} \int_A X dP. \qquad (A.32)$$

The expectation of an integrable random variable X may be conditioned on events whose outcomes are uncertain. Let's assume that $\mathcal{G}$ is a *sub-tribe* of $\mathcal{F}$, i.e. $\mathcal{G} \subset \mathcal{F}$, the conditional expectation of X, given $\mathcal{G}$, is denoted by $E[X \mid \mathcal{G}]$ and represents the expectation of X given all information contained in $\mathcal{G}$. $E[X \mid \mathcal{G}]$ is an integrable, $\mathcal{G}$-measurable random variable and satisfies:

$$\int_A X dP = \int_A E[X \mid \mathcal{G}] dP \quad \forall A \in \mathcal{G} \qquad (A.33)$$

The conditional expectation has following important properties:

i. $E[aX + bY \mid \mathcal{G}] = aE[X \mid \mathcal{G}] + bE[Y \mid \mathcal{G}]$,
   where a and b are real numbers , X and Y are integrable on $(\Omega, \mathcal{F}, P)$ and $\mathcal{G} \subset \mathcal{F}$.
ii. If $\mathcal{G}$ is the trivial σ-algebra $\{\emptyset, \Omega\}$:
   $E[X \mid \mathcal{G}] = E[X]$.
iii. If $\mathcal{G} = \mathcal{F}$, i.e. X is integrable and $\mathcal{G}$-measurable:
   $E[X \mid \mathcal{G}] = X$ almost surely (a.s.), i.e. holds at least for all subsets of Ω with non-zero probability.

iv. If Y is $\mathcal{G}$-measurable:
   $E\left[XY \mid \mathcal{G}\right] = YE\left[X \mid \mathcal{G}\right].$

v. If $\mathcal{H} \subset \mathcal{G}$:
   $E\left[E\left[X \mid \mathcal{G}\right]\mathcal{H}\right] = E\left[E\left[X \mid \mathcal{H}\right]\mathcal{G}\right] = E\left[X \mid \mathcal{H}\right].$
   This property is referred to as the *iterated conditional expectations* property or *tower* property. Iterated expectations are always equal to the expectation conditional to the smallest (coarsest) information set.

A.2.5 Stochastic Processes and Information Structure

In a dynamic setting the evolution through time of a random variable is of interest. A collection of random variables $\{X(t), t \in \mathcal{T}\}$ defined on some common probability space $(\Omega, \mathcal{F}, P)$ and assuming values in $(\Omega', \mathcal{F}')$ is called a *stochastic process*. In this book $\mathcal{T}$ always represents the time. A stochastic process is *continuous* if $\mathcal{T}$ is the real line and *discrete* if $\mathcal{T}$ evolves in integer steps. A stochastic process is termed *discrete-state process* if its values are countable. Otherwise the process is referred to a *continuous-state* process.

A stochastic process is said to be *one-dimensional* if $\Omega' = \mathbb{R}$ and $\mathcal{F}' = \mathcal{B}$ and is referred to as M-*dimensional* if $\Omega' = \mathbb{R}^M$ and $\mathcal{F}' = \mathcal{B}^M$. In the latter case the stochastic process is also termed a *vector process*.

The assignment of probability to events evolves with new information becoming available to the agents with the passage of time. Formally the information about the states of the world at time t is represented by the σ-algebra $\mathcal{F}(t)$. It is often assumed that nothing is known at the initial date and therefore $\mathcal{F}(0)$ is equal to the trivial algebra $\{\{\emptyset\}, \{\Omega\}\}$. Moreover, as time evolves new information is being gathered and it is assumed that past information is not forgotten. As a result:

$$\mathcal{F}(t) \subseteq \mathcal{F}(s), \qquad \text{for } 0 \le t < s, \tag{A.34}$$

and the non-decreasing family (or sequence in discrete-time) of σ-algebras is referred to as a *filtration* $\mathbb{F} = \{\mathcal{F}(t) : t \in \mathcal{T}\}$ and completely specifies the revelation of information through time. The filtration generated by a stochastic process is called its *natural filtration*.

If Ω is a finite set of states $\Omega = \{\omega_1, \cdots, \omega_K\}$, the finite collection of disjoint subsets of Ω whose union is Ω is called a *partition* of Ω. Formally a partition $\mathcal{P}$ of Ω is defined by:

$$\mathcal{P} = \{A'_1, \cdots, A'_N\},$$

such that:

$$A'_i \cap A'_j = \emptyset \quad \forall i, j = 1, \cdots, N \quad i \ne j, \tag{A.35}$$
$$A'_1 \cup A'_2 \cup \cdots \cup A'_N = \Omega.$$

A partition represents another way to specify the information about the states of the world. From the definitions of a σ-algebra and of a partition it is easy to verify that for a given σ-algebra on Ω there is a unique corresponding partition of Ω. For instance the partition corresponding to the trivial algebra is $\{\Omega\}$ and the partition associated with the power set yields $\{\{\omega_1\}, \{\omega_2\}, \cdots, \{\omega_K\}\}$. In a multiperiod setting the information structure is fully described by a sequence $\mathcal{P}(0), \mathcal{P}(1), \cdots, \mathcal{P}(T)$ of partitions of Ω.

A partition $\mathcal{P}$ is said to be *finer* than another partition $\mathcal{P}'$ if any event of the latter (*coarser*) partition is equal to the union of some events in the former. $\mathcal{P}$ is said to be a *refinement* of $\mathcal{P}'$, whereas $\mathcal{P}'$ is called a *coarsening* of $\mathcal{P}$. A sequence of increasing σ-algebras ("new information is incoming and nothing is forgotten") corresponds therefore to a sequence of increasingly finer partitions. From the definition of a measurable map (A.8) it results that a $\mathcal{F}$-measurable random variable X assumes constant values over each set A_i' of the partition corresponding to $\mathcal{F}$. The coarsest partition on which X is measurable is called the partition generated by X.

A stochastic process $\{X(t)\}$ is said to be *adapted* to the filtration $\mathbb{F}$ if, for any t, $X(t)$ is $\mathcal{F}(t)$-measurable. Hence, agents with the information filtration $\mathbb{F}$ will know the value of $X(t)$ at time t. Similarly, a stochastic process $\{Y(t)\}$ is said to be *predictable* or *previsible* if, for any t, $Y(t)$ is $\mathcal{F}(t-1)$ measurable.

A.2.6 Martingales

Let's consider a probability space $(\Omega, \mathcal{F}, P)$ equipped with a filtration $\mathbb{F}$. A stochastic process $\{X(t)\}$ is called a *martingale* relative to $\mathbb{F}$ and P, or more briefly a *P-martingale*, if:

i. $\{X(t)\}$ is adapted to $\mathbb{F}$,
ii. $E[X(t)] < \infty$ $t \geq 0$,
iii. $E[X(t) \mid \mathcal{F}_s] = X(s)$ (a.s.) $(0 \leq s \leq t)$.

Similarly, the process $\{X(t)\}$ is called a *super-martingale* if property (iii) becomes:

iii'. $E[X(t) \mid \mathcal{F}_s] \leq X(s)$ (a.s.) $(0 \leq s \leq t)$,

respectively a *sub-martingale* if instead of (iii):

iii''. $E[X(t) \mid \mathcal{F}_s] \geq X(s)$ (a.s.) $(0 \leq s \leq t)$,

Both super- and sub-martingales are also referred to as *semi-martingales*.

From the properties of a martingale it is obvious that the best forecast of $X(t)$, given the information set $\mathcal{F}(s)$ with $s \leq t$, is $X(s)$. A martingale with conditional expectation equals to zero is called a *fair game*:

$$E[X(t) \mid \mathcal{F}_s] = 0 \text{ (a.s.)}   (0 \leq s \leq t). \tag{A.36}$$

Clearly, the (discrete-time) process $\{Y(t)\}$ with $Y(t) = X(t) - X(t-1)$, referred to as *martingale difference*, is a fair game.

Let's consider an predictable discrete-time process $\{h(t)\}$, i.e. $\mathcal{F}(t-1)$-measurable and a discrete-time martingale $\{X(t)\}$. The adapted process $\{H(t)\}$ defined by:

$$
\begin{aligned}
H(0) &= h(0)X(0), \\
H(t) &= h(0)X(0) + \sum_{i=1}^{t} \big[h(i)\,(X(i) - X(i-1)) \big],
\end{aligned}
\tag{A.37}
$$

is called the *martingale transform* of $\{X(t)\}$ by $\{h(t)\}$. Note that the change in value $X(i) - X(i-1)$ is often denoted by $\Delta X(i)$. If $\{h(t)\}$ is bounded $\{H(t)\}$ is a martingale:

$$
\begin{aligned}
E\big[H(t) \mid \mathcal{F}(s)\big] &= E\big[H(t) - H(s) + H(s) \mid \mathcal{F}(s)\big], \\
&= E\big[h(s+1)\Delta X(s+1) + \cdots + h(t)\Delta X(t) \mid \mathcal{F}(s)\big] + H(s), \\
&\quad \text{since H(t) is adapted} \\
&= E\Big[E\big[h(s+1)\Delta X(s+1) \mid \mathcal{F}(s)\big] \mid \mathcal{F}(s)\Big] + \cdots + \\
&\quad E\Big[E\big[h(t)\Delta X(t) \mid \mathcal{F}(t-1)\big] \mid \mathcal{F}(s)\Big] + H(s), \\
&\quad \text{from the properties of the conditional expectation} \\
&= E\Big[h(s+1)E\big[\Delta X(s+1) \mid \mathcal{F}(s)\big] \mid \mathcal{F}(s)\Big] + \cdots + \\
&\quad \text{since h(t) is adapted} \\
&\quad E\Big[h(t)[E\big[\Delta X(t) \mid \mathcal{F}(t-1)\big] \mid \mathcal{F}(s)\Big] + H(s), \\
&= E\big[h(s+1)\cdot 0 \mid \mathcal{F}(s)\big] + \cdots + E\big[h(t)\cdot 0 \mid \mathcal{F}(s)\big] + H(s). \\
&\quad \text{since } \{X(t)\} \text{ is a martingale} \\
&= H(s).
\end{aligned}
\tag{A.38}
$$

A.3 Selected Distributions

A.3.1 Basic Distributions

A discrete random variable X is said to have a *Bernoulli distribution* with parameter $p \in (0,1)$, if its frequency function satisfies:

$$p_X(x) = \begin{cases} p & \text{if } x = 1, \\ 1 - p & \text{if } x = 0. \end{cases} \tag{A.39}$$

A discrete random variable X is said to have a *binomial distribution* with parameters $p \in (0, 1)$ and $n \in \mathbb{N}$, if its frequency function satisfies:

$$p_X(x) = \begin{cases} \binom{n}{x} p^x (1 - p)^{n-x} & \text{if } x = 0, 1, \cdots, n, \\ 0 & \text{elsewhere,} \end{cases} \tag{A.40}$$

where $\binom{n}{x}$ denotes the *binomial coefficients* $\frac{n!}{x!(n-x)!}$.

A discrete random variable X is said to have a *Poisson distribution* with parameter $\lambda > 0$, if its frequency function satisfies:

$$p_X(x) = \begin{cases} \frac{\lambda^x e^{-\lambda}}{x!}, & x = 0, 1, 2, \cdots \\ 0 & \text{otherwise.} \end{cases} \tag{A.41}$$

A continuous random variable X is said to have *uniform* (or *rectangular*) distribution with parameters $a > b$, if its density function satisfies:

$$f_X(x) = \begin{cases} \frac{1}{b-a}, & a \leq x \leq b \\ 0 & \text{elsewhere.} \end{cases} \tag{A.42}$$

A.3.2 Elliptically Contoured Distributions

A M-dimensional random vector X is said to be *spherically symmetric distributed* or have *spherical distribution* (SD) if:

$$OX \stackrel{d}{=} X, \tag{A.43}$$

for every $M \times M$ orthogonal matrix O, i.e. $O^{\top}O = I$ with I the $M \times M$ identity matrix. Geometrically, definition (A.43) means that spherical distributions are invariant under rotations. The characteristic function of a SD is given with:

$$\phi_X(s) = E\left[e^{is^{\top}X}\right] = \varphi(s^{\top}s) = \varphi(s_1^2 + \cdots + s_M^2), \tag{A.44}$$

for some real-valued function φ called the *characteristic generator*.

A M-dimensional random vector Y is said to be *elliptically symmetric distributed* or to have *elliptically contoured distribution* (ECD) with parameters $\mu_{(M \times 1)}$, $\Sigma_{(M \times M)}$ and φ, if:

$$Y \stackrel{d}{=} \mu + A^\top X, \tag{A.45}$$

where X has SD with characteristic function φ and $A^\top A = \Sigma$. Hence, ECD can be obtained through affine transformations of SD. The characteristic function of a ECD is can be written:

$$
\begin{aligned}
\phi_Y(s) &= E\left[e^{is^\top(\mu + A'X)}\right], \\
&= e^{(is'\mu)} E\left[e^{i(A's)'X)}\right], \\
&= e^{(is'\mu)} \varphi((A's)'(A's)), \\
&= e^{(is'\mu)} \varphi(s'\Sigma s).
\end{aligned}
\tag{A.46}
$$

Note that spherical distributions, respectively elliptically contoured distributions, are extensions of the standard M-dimensional normal distribution $\mathcal{N}(0,1)$, respectively of the M-dimensional normal distribution $\mathcal{N}(\mu, \Sigma)$. For $M = 1$, the class of ECD coincides with the class of all symmetric distributions.

Elliptically contoured random variables have a density function, if it exists, of the form:

$$f_X(x) = k_M \mid \Sigma \mid^{-\frac{1}{2}} h\left[(x_{-}\mu)^\top \Sigma(x_{-}\mu)\right], \tag{A.47}$$

where Σ is non-singular, k_M is some constant and $h[\cdot]$ is a real-valued function. Examples of ECD are the *M-dimensional normal (Gaussian) distribution* $\mathcal{N}(\mu, \Sigma)$ for which, by defining $t = (x_{-}\mu)^\top \Sigma(x_{-}\mu)$:

$$h[t] = exp(-t/2),$$

and

$$k_M = (2\pi)^{-M/2},$$

and the *M-dimensional Student-t distribution* with ν degrees of freedom where:

$$h[t] = 1 + \nu^{-1}t^{-\frac{\nu+M}{2}},$$

and

$$k_M = \frac{\Gamma\left(\frac{M+\nu}{2}\right)}{(\pi\nu)^{\frac{M}{2}} \Gamma\left(\frac{\nu}{2}\right)}.$$

The class of elliptically contoured distributions possesses following properties:

- all marginal distributions of ECD are also ECD with the same characteristic generator,

- any linear combination of ECD are also ECD with same characteristic generator,
- the density of elliptically contoured random variables is constant over the ellipsoidal contour given with following equation:

$$(x_{-}\mu)^{\top} \Sigma(x_{-}\mu) = \text{ some constant,}$$

- Σ, called the dispersion matrix, is always proportional to the covariance matrix when the variances exist.

A.3.3 Stable Distribution

A random variable X is called *stable* if for the IID variables X, X_1, X_2:

$$c_1 X_1 + c_2 X_2 \stackrel{d}{=} b(c_1, c_2)X + a(c_1, c_2), \tag{A.48}$$

for all non-negative numbers c_1, c_2 and for appropriate real numbers $b(c_1, c_2) > 0$ and $a(c_1, c_2)$.

Let S_N denote the sum $X_1 + \cdots + X_N$ of IID stable random variables. Then, following:

$$S_N \stackrel{d}{=} b_N X + c_N, \qquad N \geq 1$$
$$b_N^{-1}(S_N - c_N) \stackrel{d}{=} X, \tag{A.49}$$

holds for some real constants c_N and $b_N > 0$ and $X = X_1$.

All stable distributions have characteristic functions of the form:

$$\phi(s) = \exp\left[(i\mu s) - c \mid s \mid^\alpha \{1 + i\beta \text{ sgn } (s)z \left(\mid s \mid, \alpha\right)\}\right], \tag{A.50}$$

where $\text{sgn}(x)$ denotes the sign of x and:

$$z\left(\mid s \mid, \alpha\right) = \begin{cases} \tan\left(\frac{\pi\alpha}{2}\right) & \text{for } \alpha \neq 1, \\ \frac{2}{n}\log(s) & \text{for } \alpha = 1, \end{cases} \tag{A.51}$$

$\alpha \in (0, 2]$, $\beta \in [-1, 1]$, μ is the location parameter and c the scale parameter. If $c = 0$, the distribution is degenerate, i.e concentrated on a point. α and β are the two parameters which determine the shape of the distribution. β is a skewness parameter and therefore the symmetry of the distribution implies that $\beta = 0$. α, referred to as *characteristic exponent*, is the most important parameter and determines the moments and the tails of the distribution, as well as, as discussed below, the convergence behaviour of sums of IID variables:

- $\alpha = 1$ gives the class of *Cauchy* distributions,
- $\alpha = 2$ gives the class of normal distributions.

Note that for $\alpha = 2$, $z\,(\mid s \mid, \alpha) = 0$, since $\tan(\pi) = 0$. As a result, the characteristic function of the normal law is given with:

$$\phi(s) = \exp\left[(i\mu s) - cs^2\right], \tag{A.52}$$

and depends only on the two parameters μ and $c = \sigma^2/2$, where σ^2 is the variance. Obviously, Gaussian distributions are symmetric.

Domain of Attraction The class of the stable distributions constitutes exactly the class of all possible limit laws for (normalised and centred) sums of *IID* random variables.

A random variable X belongs to the *domain of attraction* of the α-stable distribution G_α if there exists constants $A_N \in \mathbb{R}$, $b_n > 0$ such that:

$$b_n^{-1}(S_n - A_N) \overset{d}{\to} G_\alpha. \tag{A.53}$$

A random variable X is in the domain of attraction of a normal law (stable law with $\alpha = 2$), if and only if one of the following conditions holds:

- $E[X^2] < \infty$
- $E[X^2] = \infty$ and $P\left(|X| > x\right) = 0\left(x^{-2}\int_{|Y| \leq x} y^2 dF(y)\right)$, $x \to \infty$.

The consequence of the first condition is that all distributions with finite second moment are in the domain of attraction of the normal distribution. If X is in the domain of attraction α then:

- $E[|X|^\delta] < \infty <$ for $\delta < \alpha$,
- $E[|X|^\delta] = \infty <$ for $\delta > \alpha$ and $\alpha < 2$,

i.e. $\text{var}[X] = \infty$ for $\alpha < 2$.

A.4 Types of Convergence

Convergence in distribution Let $\{X_n\}_{n=1}^\infty$ be a sequence of random variables. The sequence converges *in distribution* (*in law*) to the random variable X (written $X_n \overset{d}{\to} X$) if, for all bounded and continuous functions f the relation:

$$E[f(X_n)] \to E[f(X)], \qquad \text{as } n \to \infty \tag{A.54}$$

holds.

$X_n \overset{d}{\to} X$ holds if, and only if, for all continuity points x of the distribution function $F_X(x)$ following relation holds:

$$F_{X_n}(x) \to F_X(x), \qquad \text{as } n \to \infty. \tag{A.55}$$

If F_X is continuous, then $\{X_n\}$ converges *uniformly* to X:

$$\sup_x |F_{X_n}(x) - F_X(x)| \to 0, \qquad \text{as } n \to \infty. \tag{A.56}$$

The convergence in law is the weakest mode of probabilistic convergence as it makes statements about the distribution of random variables. Since actual values of X_n of X do not matter, the convergence in distribution may be defined in terms of probability measures and is then referred to as *weak convergence*. The modes of convergence discussed in the sequel involve values of random variables.

Convergence in Probability $\{X_n\}_{n=1}^{\infty}$ converges *in probability* to the random variable X (written $X_n \overset{P}{\to} X$) if for every positive ε:

$$P\Big(\{w \in \Omega : |X_n(w) - X(w)| > \varepsilon\}\Big) \to 0, \qquad \text{as } n \to \infty. \tag{A.57}$$

The convergence in probability implies the convergence in distribution.

Convergence in L^p Let X be a random variable such that:

$$\big(E[|X|^p]\big)^{\frac{1}{p}} < \infty, \tag{A.58}$$

for $p \geq 1$. X is said to be a pth *power integrable* random variable, which is denoted by $X \in L^p$. $\{X_n\}_{n=1}^{\infty}$ converges *in L^p*, or converges *in the pth mean*, written $X_n \overset{L^p}{\to} X$ if:

$$E\Big[|X_n - X|^p\Big] \to 0, \qquad \text{as } n \to \infty. \tag{A.59}$$

In particular, if $p = 1$, respectively $p = 2$, X_n is said to converge to X in *mean*, respectively in *mean square* or to the *quadratic mean*. If X_n converges to X in the pth mean for some p, then it implies that X_n converges to X in probability.

Almost Sure Convergence $\{X_n\}_{n=1}^{\infty}$ converges *almost surely (strongly, with probability one)* to the random variable X, written $X_n \overset{a.s.}{\to} X$, if there is an event Ω' of $\mathcal{F}$ with probability one such that for all $w \in \Omega'$:

$$X_n(\omega) \to X(\omega), \qquad \text{as } n \to \infty. \tag{A.60}$$

This equivalent to:

$$P(X_n \to X) = P(\{w \in \Omega : X_n(\omega) \to X(\omega)\}) = 1. \tag{A.61}$$

The almost sure convergence implies the convergence in probability.

A.5 Sampling Theory

Law of Large Numbers The *weak law of large numbers* in its most usual form sates that, for $X_1, \cdots, X_N$ *IID* with $E[|\, X_i\, |] < \infty$ and $E[X_i] = \mu$:

$$\bar{X}_N = \frac{1}{N} \sum_{n=1}^{N} X_n \overset{P}{\to} \mu. \tag{A.62}$$

Hence, $P(|\, \bar{X}_N - \mu\, |) \to 0$ as $N \to \infty$ and the average $\bar{X}_N$ is a consistent estimator of the mean. Note that average of uncorrelated and identically distributed random variables also converges to the quadratic mean and therefore in probability to μ if $\mathrm{var}[X_i]$ is finite (L^2 *weak law of large numbers*).

The *strong law of large numbers* asserts that for $X_1, \cdots, X_N$ pairwise independent and identically distributed, with $E[|\, X_i\, |] < \infty$ and $E[X_i] = \mu$:

$$\bar{X}_N \overset{a.s.}{\to} \mu. \tag{A.63}$$

Central Limit Theorem Let $X_1, \cdots, X_N$ be *IID* random variables in $\mathbb{R}$ with mean μ and finite variance σ^2. Then:

$$\sqrt{N}(\bar{X}_N - \mu) \overset{d}{\to} \mathcal{N}(0, \sigma^2), \tag{A.64}$$

where $\bar{X}_N = \frac{1}{N} \sum_{n=1}^{N} X_n$. Clearly, after standardisation:

$$\frac{1}{\sqrt{N}} \sum_{n=1}^{N} \frac{(X_n - \mu)}{\sigma} \overset{d}{\to} \mathcal{N}(0, 1). \tag{A.65}$$

If $X_1, \cdots, X_N$ is a sequence of *IID* random vectors in $\mathbb{R}^M$ with mean μ and covariance matrix Σ, the multivariate *central limit theorem* asserts that:

$$\sqrt{N}(\bar{X}_N - \mu) \overset{d}{\to} \mathcal{N}(0, \Sigma). \tag{A.66}$$

Berry-Esséen Theorem Note that the convergence in distribution which appears in (A.64), (A.65), and A.66) is not necessarily uniform in the underlying distribution. This implies that for a given N, the normal distribution may be a poor representation of the distribution of $\sqrt{N}(\bar{X}_N - \mu)$. However, the *Berry-Esséen theorem* shows that the convergence is uniform for distribution with finite absolute third moment $E[|\, X_\mu\, |^3]$:

$$|\, \check{F}_N(x) - \varPhi(x)\, | < \frac{cE[|\, X_\mu\, |^3]}{\sqrt{N}\sigma^3}, \tag{A.67}$$

for all x and N, $\varPhi(x)$ denotes the CDF of the standard normal distribution, c is a constant with value in $[0.4087, 0.7975]$, and $\check{F}_N$ stands for the distribution function of the normalised sum:

$$\frac{1}{\sqrt{N}} \sum_{n=1}^{N} \left(\frac{(X_n - \mu)}{\sigma} \right).$$

Large Deviations The Central Limit Theorem and its extensions, such as the Berry-Esséen theorem, give insight into the convergence of a normalised sum of IID random variables $X_1, \cdots, X_N$ to the mean, i.e. to the center of the distribution. The *theorem of large deviations* provides information about the rate of convergence of the average of a sum of IID random variables $X_1, \cdots, X_N$ to some statistics larger than the mean, i.e. located in the right tail of the distribution. Considering the value $-X_1, \cdots, -X_N$, the theorem is applicable for the left tail of the distribution. As shown in Sect. 5.3.1, the theorem of large deviations may be used to analyse the convergence of empirical estimates to a given quantile of the distribution.

Let $X_1, \cdots, X_N$ be IID, $S_N = X_1 + \cdots + X_N$ and a be $> E[X_i]$. Then, $P(S_N \geq Na) \to 0$ exponentially rapidly:

$$P(S_N \geq Na) \leq \exp\left[-N\gamma(a)\right], \quad \forall N > 0, \tag{A.68}$$

where $\gamma(a) = as - \log(\Psi(s))$ and $\Psi(s) = E[\exp(sX_i)]$ is the moment generating function of X_i. Equation (A.68) holds for arbitrary distributions, provided that $\Psi(s) < \infty$ for some $s > 0$.

The smaller upper bound for $P(S_n \geq na)$ can be found by searching the maximum of $as - \log(\Psi(s))$, that is:

$$\frac{\partial}{\partial s}\{as - \log(\Psi(s))\} = a - \frac{\Psi'(s)}{\Psi(s)} = 0. \tag{A.69}$$

Empirical Distribution Let $X_1, \cdots, X_N$ be IID random variables with distribution function $F_X(x)$. The *sample* or *empirical distribution* function $F_N(x)$ where:

$$F_N(x) = \frac{1}{N} \sum_{n=1}^{N} I_{\{X_n \leq x\}}, \quad I_{\{X_n \leq x\}} \text{ is the indicator function,} \tag{A.70}$$

represents the (non-parametrical) maximum-likelihood estimate of $F_X(x)$. Mathematically, the empirical distribution is a step function with jumps $1/N$ at $X_1, \cdots, X_N$.

Note that for a fixed x, $I_{\{X_n \leq x\}}$ is a random variable with mean $F_X(x)$. Hence, $F_N(x) \overset{a.s.}{\to} F(x)$ from the strong law of large numbers. The latter is strengthened by the *Glivenko-Cantelli theorem*:

$$\sup_x | F_N(x) - F_X(x) | \overset{a.s.}{\to} 0. \tag{A.71}$$

Bibliography

Alexander, C. (1996). Volatility and correlation forecasting. In Alexander, C., editor, *The Handbook of Risk Management and Analysis*, pages 233–260. Wiley, Chichester.

Allais, M. (1953). Le comportement de l'homme rationnel devant le risque: Critique des postulats et axiomes de l'école américaine. *Econometrica*, 21:503–546.

Antonov, I. and Saleev, V. (1979). An economic method of computing LP-sequences. *USSR Computational Mathematics and Mathematical Physics*, 19(1):252–256.

Arrow, K. (1965). *Essays in the Theory of Risk-Bearing*. North-Holland, Amsterdam.

Artzner, P., Delbaen, F., Eber, J.-M., and Heath, D. (1998). Coherent measures of risk. Preprint, University Louis Pasteur, Strasbourg.

Bachelier, L. (1900). *La théorie de la spéculation*. Gauthier-Villars, Paris.

Bawa, V. (1975). Optimal rules for ordering uncertain prospects. *Journal of Financial Economics*, 2:95–121.

Bawa, V. (1978). Safety-first, stochastic dominance, and optimal portfolio choice. *Journal of Financial and Quantitative Analysis*, 2:255–271.

Bawa, V. and Lindenberg, E. (1977). Capital market equilibrium in a mean-lower partial moment framework. *Journal of Financial Economics*, 5:189–200.

Bazaraa, M., Sherali, H., and Shetty, C. (1993). *Nonlinear Programming*. John Wiley & Sons, Inc., New York, 2 edition.

Berman, L. (1998). Accelerating Monte Carlo: quasirandom sequences and variance reduction. *Journal of Computational Finance*, 1(2):79–95.

Bernoulli, D. (1738). Specimen theoriae novae de mensura sortis. *Commentari Academiae Scientiarum Imperialis Petropolitanae*, 5(175-192). Translated as: Expositions of a New Theory on the Measurement of Risk, Econometrica 22, 1954, 23-36.

Bingham, N. and Kiesel, R. (1998). *Risk Neutral Valuation: Pricing and Hedging of Financial Derivatives*. Springer, London.

BIS (1995a). *Basle Capital Accord: treatment of potential exposure for off-balance-sheet items*. Basle Committee on Banking Supervision, Basle.

BIS (1995b). *Planned supplement to Capital Accord to incorporate market risks.* Basle Committee on Banking Supervision, Basle.

BIS (1995c). *Proposal to issue a supplement to the Basle Capital Accord to cover market risks.* Basle Committee on Banking Supervision, Basle.

BIS (1996). *Overview of the Amendment to the Capital Accord to incorporate Market Risks.* Basle Committee on Banking Supervision, Basle.

Black, F. and Scholes, M. (1973). The pricing of options of corporate liabilities. *Journal of Political Economy,* 81:637–59.

Blattberg, R. and Gonedes, N. (1974). A comparison of stable and Student distributions as statistical models for stock prices. *Journal of Business,* 47:244–280.

Bollerslev, T. (1986). Generalized autoregressive conditional heteroskedasticity. *Journal of Econometrics,* 51:307–327.

Bollerslev, T. (1987). A conditional heteroskedastic time series model for speculative prices and rates of returns. *Review of Economics and Statistics,* 69:542–547.

Bollerslev, T., Chou, R., and Kroner, K. (1992). ARCH modeling in finance. *Journal of Econometrics,* 52:5–59.

Bookstber, R. and Clarke, R. (1983a). An algorithm to calculate the return distribution of portfolios with option positions. *Management Science,* 29(4):419–429.

Bookstber, R. and Clarke, R. (1983b). *Option Strategies for Institutional Investment Management.* Addison-Wesley, Reading, Massachusetts.

Boothe, P. and Glassman, D. (1987). The statistical distribution of exchange rates. *Journal of International Economics,* 22:297–319.

Box, G. and Pierce, D. (1970). Distribution of residual autocorrelations in autoregressive-integrated moving average time series models. *Journal of the Royal Statistical Society,* B(26):211–243.

Boyle, P., Broadie, M., and Glasserman, P. (1997). Monte Carlo methods for security pricing. *Journal of Economic Dynamics and Control,* 21:1267–1321.

Bratley, P. and Fox, B. (1988). Implementing Sobol's quasirandom sequence generator. *ACM Transactions on Mathematical Software,* 14(1):88–100.

Brémaud, P. (1994). *An Introduction to Probabilistic Modeling.* Springer-Verlag, New York.

Brennan, M. and Schwartz, E. (1978). Finite difference method and jump processes arising in the pricing of contingent claims. *Journal of Financial and Quantitative Analysis,* 13:461–474.

Brockett, P. and Garven, J. (1993). A reexamination of the relationship between preferences and moment orderings by rational risk averse investors. Working paper, Univeristy of Texas, Austin.

Brooks, R. (1991). Analyzing portfolios with derivative assets: A stochastic dominance approach using numerical integration. *The Journal of Futures Markets,* 11(4):411–440.

Campbell, J., Lo, A., and MacKinlay, A. (1997). *The Econometrics of Financial Markets*. Princeton University Press, Princeton.

Cao, R., Cuevas, A., and Manteiga, W. G. (1994). A comparative study of several smoothing methods in density estimation. *Computational Statistics & Data Analysis*, 17:153–176.

Clark, P. (1973). A subordinated stochastic process model with finite variance for speculative prices. *Econometrica*, 41(1):135–155.

Copeland, T. and Weston, J. (1992). *Financial Theory and Corporate Policy*. Addison-Wesley, Reading, 3 edition.

Cox, J., Ingersoll, J., and Ross, S. (1981). The relation between forward prices and futures prices. *Journal of Financial Economics*, 9:321–46.

Cox, J., Ross, S., and Rubinstein, M. (1979). Option pricing: a simplified approach. *Journal of Financial Economics*, 7:229–264.

Cox, J. and Rubinstein, M. (1985). *Options Markets*. Prentice Hall, Englewood Cliffs.

Davies, R. (1973). Numerical inversion of a characteristic function. *Biometrika*, 60(2):415–417.

Davies, R. (1980). The distribution of a linear combination of chi-square random variables. *Applied Statistics*, 29:323–333.

Dielman, T., Lowry, C., and Pfaffenberger, R. (1994). A comparison of quantile estimators. *Communications in Statistics : Simulation*, 23(2):355–371.

Duffie, D. (1996). *Dynamic Asset Pricing Theory*. Princeton University Press, Princeton, 2 edition.

Duffie, D. and Pan, J. (1997). An overview of Value-at-Risk. *The Journal of Derivatives*, 4(3):7–49.

Durrett, R. (1996). *Probability Theory and Examples*. Duxbury Press, Belmont, 2 edition.

Efron, B. and Tibshirani, R. (1993). *An Introduction to the Bootstrap*. Chapman & Hall, New York.

Elton, E. and Gruber, M. (1995). *Modern Portfolio Theory and Investment Analysis*. Wiley, New York, 5 edition.

Embrechts, P., Klüppelberg, C., and Mikosch, T. (1997). *Modelling Extremal Events*. Springer, Berlin.

Embrechts, P., McNeil, A., and Straumann, D. (1998). Correlation and dependency in risk management: Properties and pitfalls. Working paper, Swiss Federal Institute of Technology, Zürich.

Engle, R. (1982). Autoregressive conditional heteroscedasticity with estimates of the variance of United Kingdom inflation. *Econometrica*, 50(4):987–1007.

Estrella, A. (1995). Taylor, Black and Scholes: Series approximations and risk management pitfalls. Research Paper 9501, Federal Reserve of New York, New York.

Fabozzi, F. (1996). *Bond Markets, Analysis and Strategies*. Prentice Hall, Upper Saddle River.

Fama, E. (1965). The behavior of stock market prices. *Journal of Business*, 38:34–105.

Fama, E. (1970). Efficient capital markets: A review of theory and empirical work. *Journal of Finance*, 25(2):383–417.

Fama, E. (1976). *Foundations of finance*. Basic Books, New York.

Fang, K., Kotz, S., and Ng, K. (1990). *Symmetric Multivariate and Related Distributions*. Chapman and Hall, London.

Faure, H. (1982). Discrépance de suites associées à un système de numération (en dimension s). *Acta Arithmetica*, XLI:337–351.

Feller, W. (1966). *An Introduction to Probability Theory*, volume 2. Wiley & Sons, New York.

Ferguson, T. (1996). *A Course in Large Sample Theory*. Chapman & Hall, London.

Fishburn, P. (1970). *Utility Theory for Decision Making*. Wiley & Sons, New York.

Fishburn, P. (1977). Mean-risk analysis with risk associated with below-target returns. *The American Economic Review*, 67(2):116–126.

Fisher, I. (1930). *The theory of Interest Rates ,reprinted in: "The Works of Irving Fisher", 1997*, volume 9. ed. W.J. Barber, Pickering & Chatto, London.

Fishman, G. (1996). *Monte Carlo : Concepts, Algorithms and Applications*. Springer, New York.

Fox, B. (1986). Implementation and relative efficiency of quasirandom sequence generators. *ACM Transactions on Mathematical Software*, 12(4):362–376.

Fraleigh, J. and Beauregard, R. (1995). *Linear Algebra*. Addison Wesley, Reading, 3 edition.

Frauendorfer, K. (1992). *Stochastic Two-Stage Programming*. Lecture Notes in Economics and Mathematical Systems 392. Springer-Verlag, Berlin.

Frauendorfer, K. (1996). Barycentric scenario trees in convex multistage stochastic programming. *Mathematical Programming*, 75(2):277–294.

Frauendorfer, K. and Härtel, F. (1996). On the goodness of discretizing stochastic processes in stochastic multistage programming. Working paper, Institute of Operations Research, University of St. Gallen.

Frauendorfer, K. and Königsperger, E. (1995). Approximation of p&l distributions. *Risklab Report, University of St. Gallen*.

Frauendorfer, K. and Marohn, C. (1996). Refinement issues in stochastic multistage linear programming. Working paper, Institute of Operations Research, University of St. Gallen.

Frauendorfer, K., Moix, P.-Y., and Schmid, O. (1997). Approximation of profit-and-loss distributions (part II). *Risklab Report, University of St. Gallen*.

French, K. (1980). Stock returns and the weekend effect. *Journal of Financial Economics*, 8:55–70.

Friend, I. and Blume, M. (1975). The demand for risky assets. *American Economic Review*, pages 900–922.

Froot, K., Scharfstein, D., and Stein, J. (1994). A framework for risk management. *Harvard Business Review*, pages 91–102.

Fuller, W. (1976). *Introduction to Statistical Times Series*. Wiley & Sons, New York.

Galanti, S. and Jung, A. (1997). Low-discrepancy sequences: Monte Carlo simulation of option prices. *The Journal of Derivatives*, pages 63–83.

Gastineau, G. (1992). *Dictionary of Financial Risk Management*. Probus Publishing Company, Chicago.

Gélinas, R. (1984). *Suites et Séries et Transformées de Fourier; Variables Complexes*. Editions SMG, Trois-Rivières.

Gibbons, J. (1971). *Nonparametric Statistical Inference*. McGraw-Hill, New York.

Gibbons, J. and Chakraborti, S. (1992). *Nonparametric Statistical Inference*. Marcel Dekker, New York, 3 edition.

Gil-Pelaez, J. (1951). Note on the inversion theorem. *Biometrika*, pages 481–482.

Granger, C. and Morgenstern, O. (1970). *Predicability of Stock Market Prices*. Heath, Lexington.

Granger, C. and Newbold, P. (1986). *Forecasting Economic Time Series*. Academic Press, San Diego.

Hadar, J. and Russel, W. (1969). Rules for ordering uncertain prospects. *American Economic Review*, 59:167–197.

Hall, P. and Marron, J. (1987). Estimation of integrated squared density derivatives. *Statistics and Probability Letters*, 6:109–115.

Hall, P. and Marron, J. (1991). Lower bounds for bandwidth selection in density estimation. *Probability Theory and Related Fields*, 90:149–173.

Hall, P., Sheather, S., Jones, M., and Marron, J. (1991). On optimal data-based bandwidth selection in kernel density estimation. *Biometrika*, 78:263–269.

Halton, J. (1960). On the efficiency of cetain quasi-random sequences of points in evaluating multi-dimensional integrals. *Numerical Methematics*, 2:84–90.

Hamilton, J. D. (1994). *Time Series Analysis*. Princeton University Press, Princeton.

Hanoch, G. and Levy, H. (1969). The efficiency analysis of choices involving risk. *Review of Economic Studies*, pages 335–346.

Härdle, W. (1991). *Smoothing Techniques: with Implementation in S*. Springer, New York.

Harrell, F. and Davis, C. (1982). A new distribution-free quantile estimator. *Biometrika*, 69(3):635–640.

Harrison, J. and Kreps, D. (1979). Martingales and arbitrage in multiperiod securities markets. *Journal of Economic Theory*, 20:381–408.

Harrison, J. and Pliska, S. (1981). Martingales and stochastic integrals in the theory of continuous trading. *Stochastic Processes and their Applications*, 11:215–260.

Harvey, A., Ruiz, E., and Shephard, N. (1994). Multivariate stochastic variance models. *Review of Economic Studies*, 61:247–264.

Hendricks, D. (1996). Evaluation of Value-at-Risk models using historical data. *FRBNY Economic Policy Review*.

Heuser, H. (1993). *Lehrbuch der Analysis*, volume 2. B.G. Teubner, Stuttgart, 8 edition.

Hodges, J. and Lehmann, E. (1956). The efficiency of some nonparametric competitors of the t-test. *Annals of Mathematical Statistics*, 27:324–335.

Huang, C. and Litzenberger, R. (1988). *Foundations for Financial Economics*. Elsvier, New York.

Huissman, R., Koedijk, K., Kool, C., and Palm, F. (1998). The fat-tailedness of FX returns. Working paper, LIFE, Maastricht University, Maastricht.

Hull, J. (2000). *Options, Futures, and Other Derivatives*. Prentice Hall, Upper Saddle River, 4 edition.

Hull, J. and White, A. (1988). The use of the control variate technique in option pricing. *Journal of Financial and Quantitative Analysis*, 23(3):237–251.

Hurst, S. and Platen, E. (1997). The marginal distributions of returns and volatility. In Dodge, Y., editor, *L1-Statistical Procedures and Related Topics*, volume 31 of *IMS Lecture Notes-Monograph Series*, pages 301–314. Institute of Mathematical Statistics, Hayward.

Imhoff, J. (1961). Computing the distribution of qudratic forms in normal variables. *Biometrika*, 48:419–426.

Ingersoll, J. (1987). *Theory of Financial Decision Making*. Rowman & Littlefield, Savage.

Jarrow, R. and Rudd, A. (1983). *Option pricing*. Richard D. Irwin, Homewood.

Jobson, J. (1991). *Applied Multivariate Data Analysis: Regression and Experimental Design*, volume 1. Springer Verlag, New York.

Johnson, M. (1987). *Multivariate Statistical Simulation*. Wiley & Sons, New York.

Johnson, N., Kotz, S., and Balakrishnan, N. (1994). *Continuous Univariate Distributions*, volume 1. Wiley & Sons, New York, 2 edition.

Johnson, N. L. and Kotz, S. (1970). *Continuous Univariate Distributions*, volume 2. Houghton Mifflin Company, Boston, 1 edition.

Jones, M., Marron, J., and Park, B.-U. (1991). A simple root n bandwidth selector. *Annals of Statistics*, 19:1919–1932.

Jones, M., Marron, J., and Sheather, S. (1996a). A brief survey of bandwidth selection for density estimation. *Journal of the American Statistical Association*, 91(433):401–407.

Jones, M., Marron, J., and Sheather, S. (1996b). Progress in data-based bandwidth selection for kernel density estimation. *Computational Statistics*, 11:337–381.

Jorion, P. (1995). Predicting volatility in the foreign exchange market. *The Journal of Finance*, 50:507–528.

Joy, C., Boyle, P., and Tan, K. (1996). Quasi-Monte Carlo methods in numerical finance. *Management Science*, 42(6).

JP Morgan (1996). *RiskMetrics - Technical Document*. J.P Morgan.

Kahneman, D. and Tversky, A. (1979). Prospect theory: an analysis of decision under risk. *Econometrica*, 47:263–291.

Kaigh, W. and Lachenbruch, P. (1982). A generalized quantile estimator. *Communications in Statistics: Theory and Methods*, 11(19):2217–2238.

Kalos, M. H. and Whitlock, P. (1986). *Monte Carlo Methods*, volume 1. John Wiley & Sons, New York.

Kim, D. and Kon, S. (1994). Alternative models for the conditional heteroscedasticity of stock returns. *Journal of Business*, 67(4):563–598.

Kincaid, D. and Cheney, W. (1996). *Numerical Analysis*. Brooks/Cole Publishing Company, Pacific Grove, 2 edition.

Knuth, D. (1998). *The Art of Computer Programming: Seminumerical Algorithms*, volume 2. Addison Wesley, Reading, 3 edition.

Kolmogorov, A. (1933). *Grundbegriffe der Wahrscheinlichkeitsrechnung*. Springer, Berlin.

Kon, S. (1984). Models of stock returns: A comparison. *The Journal of Finance*, 39(1):147–1633.

Kreps, D. (1988). *Notes on the Theory of Choice*. Westview Press, Boulder.

Lehmer, D. (1951). Mathematical methods in large-scale computing. *Annu. Comput. Lab.*, 26:141–146.

LeRoy, S. (1989). Efficient capital markets and martingales. *Journal of Economic Literature*, 27:1583–1621.

Levy, H. (1969). A utility function depending on the first three moments. *Journal of Finance*, 24(4):715–719.

Lintner, J. (1965). Security prices, risk, and maximal gains from diversification. *Journal of Finance*, pages 587–615.

Ljung, G. and Box, G. (1978). On a measure of lack of fit in times series models. *Biometrika*, 66:67–72.

Lo, A. (1999). The three P's of total risk management. *Financial Analyst Journal*, 55(1):13–26.

Loomes, G. and Sugden, R. (1982). Regret theory: An alternative theory of rational choice under uncertainty. *The Economic Journal*, 92:805–824.

Mandelbrot, B. (1963a). New methods in statistical economics. *Journal of Political Economy*, 71(421-440).

Mandelbrot, B. (1963b). The variation of certain speculative prices. *Journal of Business*, 36:394–419.

Mandelbrot, B. (1967). The variation of some other speculative prices. *Journal of Business*, 40:393–413.

Manoukian, E. (1986). *Modern Concepts and Theorems of Mathematical Statistics*. Springer, New York.

Markowitz, H. (1952). Portfolio selection. *Journal of Finance*, 7:77–91.

Marohn, C. (1998). *Stochastische mehrstufige lineare Programmierung im Asset & Liability Management*. PhD thesis, University of St. Gallen, Haupt, Bern.

Marsaglia, G. (1968). Random numbers fall mainly in the planes. *Proceedings of the National Academy of Sciences*, 61:25–28.

Mas-Colell, A., Whinston, M., and Green, J. (1995). *Microeconomic Theory*. Oxford University Press, New York.

Merton, R. (1992). *Continuous-Time Finance*. Blackwell, Cambridge, revised edition.

Modigliani, F. and Miller, M. (1958). The cost of capital, corporation finance and the theory of investment. *American Economic Review*, 48:261–97.

Moro, B. (1995). The full Monte. *RISK*, 8(2):57–58.

Mossin, J. (1966). Equilibrium in a capital asset market. *Econometrica*, pages 768–783.

Neftci, S. (1996). *Mathematics of Financial Derivatives*. Academic Press, San Diego.

Niederreiter, H. (1992). *Random Number and Quasi-Monte Carlo Methods*. Society for Industrial and Applied Mathematics, Philadelphia.

Norton, C. and Blume, L. (1994). *Mathematics for Economists*. Norton, New York.

Ökten, G. (1998). Error estimation for quasi-monte carlo methods. In Niederreiter, H., Hellekalek, P., Larcher, G., and Zinterhof, P., editors, *Monte Carlo and Quasi-Monte Carlo Methods 1996*, pages 353–367. Springer, New York.

Osborne, M. (1959). Brownian motion in the stock market. *Operations Research*, 7:145–173.

Park, B.-U. and Marron, J. (1990). Comparison of data-driven bandwidth selectors. *Journal of the American Statistical Association*, 85:66–72.

Parrish, R. (1990). Comparison of quantile estimators in normal sampling. *Biometrics*, 46:247–257.

Parzen, E. (1962). On estimation of a probability density function and mode. *Annals of Mathematical Statistics*, 33:1065–1076.

Paskov, S. (1997). New methodolgies for valuing derivatives. In Dempster, M. and Pliska, S., editors, *Mathematics of Derivative Securities*, chapter 27, pages 545–582. Cambridge University Press, Cambridge.

Pelsser, A. and Vorst, T. (1996). Optimal optioned portfolios with confidence limits on shortfall constraints. Report 9604, Erasmus University, Rotterdam.

Platt, R. (1986). *Controlling Interest Rate Risk*. Wiley & Sons, New York.

Pliska, S. (1997). *Introduction to Mathematical Finance*. Blackwell, Malden Ma.

Praetz, P. (1972). The distribution of share price changes. *Journal of Business*, 45:49–55.

Pratt, J. (1964). Risk aversion in the small and in the large. *Econometrica*, 32:122–136.

Press, S. (1968). A compound events model for security prices. *Journal of Business*, 41:317–35.

Press, W. and Teukolsky, S. (1989). Sobol algorithm. *Numerical recipes Column*, 3(6).

Press, W., Vetterling, W., Teukolsky, S., and Flannery, B. (1992). *Numerical Recipes in C*. Cambridge University Press, Cambridge, 2 edition.

Rao, C. (1973). *Linear Statistical Inference and Its Applications*. Wiley & Sons, New York, 2 edition.

Rebonato, R. (1998). *Interest-Rate Option Models*. Wiley & Sons, Chichester, 2 edition.

Reilly, F. (1994). *Investment Analysis and Portfolio Management*. The Dryden Press, Orlando.

Reiss, R. (1989). *Approximate Distributions of Order Statistics*. Springer, New York.

Rice, J. (1995). *Mathematical Statistics and Data Analysis*. Duxbury Press, Belmont, 2 edition.

Richter, M. (1960). Cardinal utility, portfolio selection and taxation. *Review of Economic Studies*.

Ripley, B. (1987). *Stochastic Simualtion*. John Wiley & Sons, New York.

Roger, P. (1991). *Les Outils de la Modélisation Financière*. Presses Universitaires de France, Paris.

Ross, S. (1976). The arbitrage theory of capital asset pricing. *Journal of Economic Theory*, 13(3).

Ross, S. (1990). *A Course in Simulation*. Macmillan Publishing company, New York.

Rouvinez, C. (1997). Going greek with VaR. *RISK*, 10(2).

Ruiz, E. (1994). Quasi-maximum likelihood estimation of stochastic volatility models. *Journal of Econometrics*, 63:289–306.

Samuelson, P. (1965). Proof that properly anticipated prices fluctuate randomly. *Industrial Management Review*, 6:41–49.

Savage, L. (1954). *The Foundations of Statistics*. Wiley & Sons, New York.

Schervish, M. (1995). *Theory of Statistics*. Springer, New York.

Schiller, R. (1997). Human behavior and the efficiency of the financial system. preprint.

Schwartz, E. (1977). The valuation of warrants: Implementing a new approach. *Journal of Financial Economics*, 4:79–94.

Scott, D. and Terrell, G. (1987). Biased and unbiased cross-validation in density estimation. *Journal of the American Statistical Association*, 82:1131–1146.

Sharpe, W. (1964). Capital asset prices: A theory of market equilibrium under conditions of risk. *Journal of Finance*, pages 425–442.

Sharpe, W. (1978). *Investments*. Prentice Hall, Englewood Cliff.

Sheather, S. (1992). The performance of six popular bandwidth selection methods on some real data sets. *Computational Statistics*, 7:225–250.

Sheather, S. and Jones, M. (1991). A reliable data-bandwidth selection method for kernel density estimation. *Journal of the Royal Statistical Society, Series B*, 53:683–690.

Shimko, D. (1992). *Finance in Continuous Time: A Primer*. Kolb Publishing Company, Miami.

Silverman, B. (1986). *Density Estimation for Statistics and Data Analysis*. Chapman and Hall, London.

Sobol, I. (1976). Uniformly distributed sequences with an additional uniform property. *USSR Comput. Math. Math. Phys.*, 14:236–242.

Sobol, I. (1979). On the systematic search in a hypercube. *SIAM Journal of Numerical Analysis*, 16:790–793.

Spremann, K. (1992). Probleme mit risiken. *Schweizer Bank*, 10:87–89.

Spremann, K. (1996). *Wirtschaft, Investition und Finanizierung*. R. Oldenburg, 5 edition.

Springer, M. (1979). *The Algebra of Random Variables*. Johne Wiley & Sons, New York.

Stoyan, D. (1983). *Comparison Methods for Queues and Other Stochastic Models*. Wiley & Sons.

Studer, G. (1997). *Maximum Loss for Measurement of Market Risk*. PhD thesis, Swiss Federal Institute of Technology, Zurich.

Sundaram, R. (1997). Equivalent martnigale measures and risk neutral pricing: An expository note. *The Journal of Derivatives*, pages 85–98.

Takayama, A. (1994). *Analytical Methods in Economics*. Harvester Wheatsheaf, Hertfordshire.

Taylor, S. (1986). *Modelling Financial Times Series*. Wiley & Sons, Chichester.

Tucker, A. (1992). A reexamination of finite- and infinite-variance distributions as models of daily stock returns. *Journal of Business & Economic Statistics*, 10:73–81.

Tucker, L. and Pond, L. (1988). The probability distribution of foreign exchanges: Tests of candidate processes. *The Review of Economics and Statistics*, 70(4):638–647.

Tversky, A. and Kahneman, D. (1992). Advances in prospect theory: Cumulative representation of uncertainty. *Journal of Risk and Uncertainty*, 5:297–323.

Venables, W. and Ripley, B. (1994). *Modern Applied Statistics with S-Plus*. Springer-Verlag, New York.

von Neumann, J. and Morgenstern, O. (1944). *Theory of Games and Economic Behaviour*. Princeton University Press, Princeton.

von Neumann, J. and Morgenstern, O. (1947). *Theory of Games and Economic Behaviour*. Princeton University Press, Princeton, 2 edition.

Wang, S. (1997). Aggregation of correlated risk portfolios: Models & algorithms. Working paper, CAS Committee on Theory of Risk.

Wilmott, P., Dewynne, J., and Howison, S. (1994). *Option Pricing*. Oxford Financial Press, Oxford.

Wonnacott, T. and Wonnacott, R. (1990). *Introductory Statistics*. Wiley & Sons, New York.

List of Figures

List of Tables

Index

Lecture Notes in Economics
and Mathematical Systems

For information about Vols. 1–315
please contact your bookseller or Springer-Verlag